# LIGHT EMITTING DIODES
## AN INTRODUCTION

# LIGHT EMITTING DIODES
## AN INTRODUCTION

### Klaus Gillessen
### Werner Schairer
Telefunken Electronic, West Germany

Prentice/Hall International

Englewood Cliffs, N.J.   London   Mexico   New Delhi   Rio de Janeiro
Singapore   Sydney   Tokyo   Toronto

**British Library Cataloguing in Publication Data**

Gillessen, Klaus
Light emitting diodes: an introduction. –
(Prentice-Hall International series in optoelectronics)
1. Light emitting diodes
I. Title  II. Schairer, Werner
621.3815′22       TK7871.89.L53

ISBN 0-13-536533-3

Prentice-Hall Inc., *Englewood Cliffs, New Jersey*
Prentice-Hall International (UK) Ltd, *London*
Prentice-Hall of Australia Pty Ltd, *Sydney*
Prentice-Hall Canada Inc., *Toronto*
Prentice-Hall Hispanoamericana S.A., *Mexico*
Prentice-Hall of India Private Ltd, *New Delhi*
Prentice-Hall of Japan Inc., *Tokyo*
Prentice-Hall of Southeast Asia Pte Ltd, *Singapore*
Editora Prentice-Hall do Brasil Ltda, *Rio de Janeiro*

Printed and bound in Great Britain for
Prentice-Hall International (UK) Ltd,
66 Wood Lane End, Hemel Hempstead, Hertfordshire, HP2 4RG
at the University Press, Cambridge.

1 2 3 4 5   91 90 89 88 87

ISBN 0-13-536533-3

# CONTENTS

# PREFACE

Light emitting diodes (LEDs) are - in contrast to incandes-
cent lamps - cold light sources. Cold light emission means
that the temperature of the source is far from equilibrium
with the emitted radiation. Therefore non-equilibrium has to
be achieved. The key to this is charge carrier injection
across a pn junction within a semiconductor crystal into a
zone where the injected carriers can convert their excess
energy into light. This process is the principle of light
emission in diodes.

LEDs acquired high importance within a relatively short time.
Starting from humble beginnings, their brightness was contin-
uosly stengthened. As a consequence, LEDs became able to con-
quer quite a strong market position: today light emitting
diodes are produced in very large numbers: estimations for
the year 1986 yield worldwide production numbers in the order
of magnitude of $5 \times 10^9$. The turnover achieved with LEDs and
LED devices alone is estimated to be around 200 millions of
dollars, whereas the turnover for systems employing LEDs is
certainly several orders of magnitude larger. The chemical
element gallium, as an essential component of most LEDs, is
heavily consumed for LED production. In effect it forms the
major part of the world gallium market, which was about 60
tons in the year 1985.

As can be seen from these market figures, LEDs are products
with an increasing degree of maturity. This does not mean
that LED development has come to an end. On the contrary, new
types are introduced and new applications are opened up per-
manently. In this situation it seemed appropriate to try a
comprehensive presentation of the subject.

Because LEDs are semiconductor devices, technological aspects
are just as important as physical principles. This book tries
to give both sides appropriate emphasis. While there are excellent texts covering the physics behind the devices, a con-
sistent presentation of LED technology is missing. Therefore
we thought it important to include in this book technological
methods and problems.

The book should be useful for undergraduate students who wish
to acquire knowledge in a particular field of optoelectronics, either by studying the book alone or by using it as an
accompaniment to an optoelectronics course. But also profes-
sionals intending to start developmental work on LEDs, or to
produce or use LEDs, should find this introduction helpful.
The authors hope that even LED specialists will find something new for them here.

The chapters follow a quite natural order:

- The first chapter is a brief summary or repetition of some
  selected parts of solid state and semiconductor physics
  which are required to understand other parts of the book.
  Readers who feel strong in this field may skip chapter 1.

- In the second chapter LED materials are treated which are
  mostly III-V compounds, with the largest importance at-
  tached to gallium arsenide (GaAs), gallium phosphide (GaP),
  gallium arsenide phosphide ($GaAs_{1-x}P_x$), and gallium alumi-

num arsenide $(Ga_{1-x}Al_xAs)$.

- The third chapter gives an overview of LED technology, that
  is the sum of techniques required for fabrication of LED
  chips, in particular crystal growth and epitaxy of the respective III-V compounds, and specific wafer processing
  steps.

- The structures of the most common LED chip types are described in the fourth chapter, including visible LEDs of
  all colors, and infrared emitting types for standard and
  specialized purposes, e.g. optical fiber transmission.

- The fifth chapter gives an introduction into the field of
  optical measurements. The commonly used units are defined,
  the problem of color characterization is treated, and measuring techniques are described.

- In the sixth chapter the optoelectronic properties of LEDs
  are dealt with. The most important term here is efficiency
  which determines the brightness of an LED and the optical
  power of an IRED. Furthermore, the spectral properties of
  LEDs are given which determine the color of visible LEDs.
  This chapter ends with a section on reliability of LEDs,
  also summarizing relevant degradation mechanisms.

- The seventh and last chapter deals with application of LEDs
  and IREDs. Some general rules for the applicability of LEDs
  are derived from the specific properties of the devices.
  Regions are also defined where the application of LEDs is
  not sensible. The applications of visible LEDs are treated
  in the order of increasing complexity, from single ON-OFF
  indicators to experimental flat screens. Applications of
  IREDs include remote control, optocouplers, and optical
  communication.

Each chapter is provided with questions and references. The
questions will generally help the student to test his under-
standing of the text, but a few give at the same time deeper
insight into matters dealt with in the preceeding chapter.
Carefully selected references, although few in number, offer
the opportunity to enter into special fields which are not
covered adequately in the text. Some references have also
been included because of their importance in the historical
development of LEDs.

We would like to thank our employer, TELEFUNKEN electronic
GmbH, for the use of its facilities in preparing the manu-
script. In particular we are grateful to Miss S. Lier who
typed the text patiently in several states of perfection. We
also want to express our gratitude to Mrs. R. Höfer, Mrs.
A. Müller, and Mr. D. Wallis, who prepared the figures. Finally, we thank many of our colleagues for helpful discus-
sions.

                                   K. Gillessen and W. Schairer

# 1 PHYSICAL BACKGROUND

This chapter is not intended to replace a textbook on solid state physics. Instead, its aim is to bring some required basic knowledge back into the mind of the reader, who is assumed already to know something about the subject. For this purpose we will present a short overview of the particular parts of semiconductor physics which are necessary to understand light emitting diodes. Of course some background in general physics is also required and cannot be repeated here.

## 1-1 Semiconductor crystals

Crystals are bodies of condensed matter with ordered structures, which means that the atoms which form a crystal are set up in a regular manner. In crystallography these structures are described with geometrical models called the lattices. From the large number of lattice types which are realized in nature we will only mention the few examples which are important for semiconductors. In particular, the following ideas apply to crystals with cubic symmetry but are not neccessarily true for all crystals. This is not a serious restriction, because most semiconductors crystallize in cubic lattices.

Directions and planes in a crystal lattice are described using the Miller indices  h k l , which are placed in various types of bracket:

$\begin{bmatrix} h & k & 1 \end{bmatrix}$ means the direction with the components h along the x-axis, k along the y-axis, and 1 along the z-axis. For example $\begin{bmatrix} 110 \end{bmatrix}$ is the direction in the x-y plane which divides the angle between the x- and y-axes into two equal parts.

$\langle h\ k\ 1 \rangle$ means all equivalent directions, i.e. all directions given by exchanging h, k, 1 and changing to negative values $\bar{h}$, $\bar{k}$, $\bar{1}$. For example, $\begin{bmatrix} 110 \end{bmatrix}$, $\begin{bmatrix} 011 \end{bmatrix}$, $\begin{bmatrix} \bar{1}10 \end{bmatrix}$ are equivalent.

$\left( h\ k\ 1 \right)$ describes a plane which cuts the axes at $h^{-1}$, $k^{-1}$, $1^{-1}$. For example, the (100) plane does not cut the y and z axes ($k^{-1} = \infty$, $1^{-1} = \infty$), therefore it is parallel to the y-z plane.

$\{ h\ k\ 1 \}$ describes all equivalent planes. For example, (111), ($\bar{1}$11), (1$\bar{1}\bar{1}$) and so on are equivalent planes.

In cubic crystals the three axes are perpendicular to each other and all directions $\begin{bmatrix} h & k & 1 \end{bmatrix}$ are normal to the planes with the same Miller indices (h k 1). Some low index directions and planes for the cubic lattice are shown in Fig. 1-1. The cube in Fig. 1-1 is called the unit cell of the cubic lattice. To build a crystal lattice, the unit cell is repeated in all directions. The side length of the unit cell is called the lattice constant a of the cubic lattice.

If one atom is placed on each corner of the unit cell, the simple cubic lattice results. An example of this structure is sodium chloride, NaCl. If atoms are located on the corners and on the centers of the cube faces, the resulting structure is called face-centered cubic (fcc). Assuming the atoms to be spheres touching each other, the fcc lattice achieves the

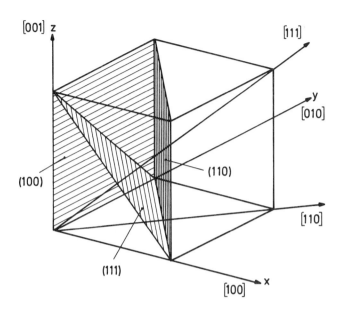

Fig. 1-1     Low index directions and planes in cubic
             lattices

densest possible packing of spheres. This structure is realized with many metals, for example, aluminum, copper, silver,
gold. From the face-centered cubic it is only a small step to
the diamond lattice, which is found with diamond and the elemental semiconductors silicon, germanium and (gray) tin. The
diamond lattice can be visualized as two fcc lattices which
are displaced against each other along the $\left[111\right]$ direction
(the body diagonal of the unit cube), so that the origin of
the second lattice has the coordinates 1/4, 1/4, 1/4. This is
shown in Fig. 1-2, where the atoms of the first fcc lattice
are indicated by the symbols o , and the atoms of the second
fcc lattice by + . The chemical bonds between nearest neighbours are depicted with heavy lines. There are 8 atoms per
unit cell: 8 atoms at the cube corners, which count 1/8 each
only, because they belong to 8 adjacent unit cells ($\rightarrow$ 1 atom/

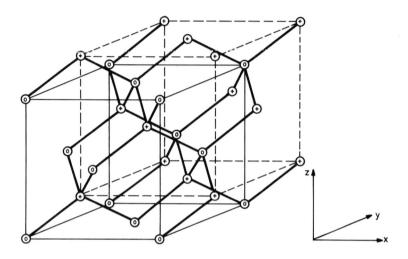

Fig. 1-2    The diamond lattice consisting of two fcc
            sublattices which are displaced along
            [111]

unit cell), 6 at the cube faces, counting 1/2 each (→ 3 atoms/
unit cell), and 4 atoms in the volume of the cube. The distance between nearest neighbours is the body diagonal of a
cube with side length a/4, which amounts to a $\sqrt{3}/4$.

Many compound semiconductors, which are described in more detail in chapter 2, have the zincblende structure, which is
closely related to the diamond lattice. In the zincblende
lattice the atoms belonging to one fcc sublattice are of one
type, and the atoms of the other sublattice are of another
type. This is also included in Fig. 1-2, if the different
symbols are interpreted as different atoms. Although diamond
and zincblende are very similar, there is one important difference: the zincblende structure is not symmetrical along

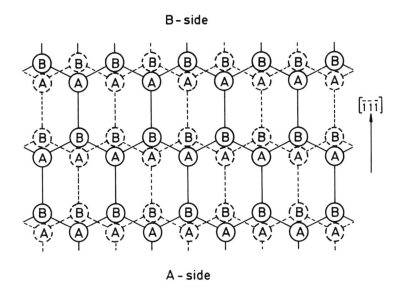

B-side

A-side

Fig. 1-3    Zincblende lattice showing double layers
            of A and B atoms in (111) planes

the $\langle 111 \rangle$ directions. This is shown in Fig. 1-3, which is a
view of a zincblende crystal with the $[111]$ direction in the
plane of the figure. In this perspective it can be seen that
the crystal consists of double layers containing A and B
atoms, so that one face of the crystal, the ($\overline{1}\overline{1}\overline{1}$) face, has
only B atoms at the surface, whereas the opposite (111) face
has only A atoms. As one might imagine, the A and B sides
differ considerably in their physical and chemical properties.

Other compound semiconductors crystallize in the wurtzite
lattice, which is similar to the zincblende lattice. In fact,
the relation between nearest neighbours is identical with the
two lattices, they differ only in the stacking sequence of
the double layers, as already depicted in Fig. 1-3. The dif-

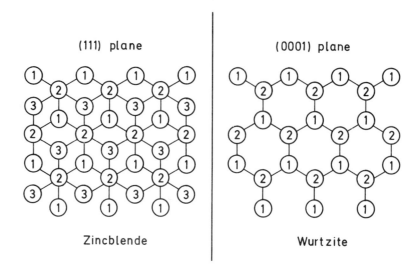

(111) plane                         (0001) plane

Zincblende                          Wurtzite

Fig. 1-4    Stacking sequence of layers in the zincblende and wurtzite lattices

ference in stacking is shown schematically in Fig. 1-4. In
the zincblende lattice there are three different stacking
positions, so that the fourth double layer is in the same
position as the first. The wurtzite lattice has only two different stacking positions, and the third double layer is exactly on top of the first layer. The wurtzite structure is
hexagonal, a class of lattices with a lower degree of symmetry, which is usually described with a set of four Miller indices. The (0001) plane of the wurtzite lattice corresponds
to the (111) plane of the zincblende lattice.

Up to now we have dealt with examples of ideal crystal lattices. In practice, however, crystals are not ideal; they always contain defects. Crystal defects are usually classified
according to their dimensions:

-   zero-dimensional (point) defects:
    intrinsic defects: vacanies, interstitials,
                             antisite defects
    extrinsic defects: impurity atoms

-   one-dimensional (line) defects:
    dislocations (screw or step type)

-   two-dimensional defects:
    grain boundaries
    stacking faults
    crystal surface

-   three-dimensional defects:
    inclusions of foreign matter

Impurities are frequently introduced intentionally into semiconductor crystals; this is known as doping. Unintentionally
incorporated impurities and all other defects, however, may
exert an unwanted influence on the properties of the crystal.
Examples will be discussed in several chapters of this book.


## 1-2 Direct and indirect semiconductors

The electrons of an atom can only occupy discrete energy levels, whereas in solids the allowed states form certain energy
ranges, the so-called energy bands. At very low temperatures
all states up to the Fermi level are occupied, and all higher
states are empty. With increasing temperature the transition
between occupied and empty states becomes more and more gradual. The position of the Fermi level relative to the energy
bands determines very much the electrical properties of a
solid: if the Fermi level is located within an energy band,
the solid has metallic properties, if it falls between two
bands, the solid can be a semiconductor or an insulator. The

energy difference between the highest occupied and the lowest
empty band is called the band gap $E_g$. Band gaps are usually
given in electron volts, 1 eV = 1.6022 x $10^{-19}$ J. If $E_g$ is
larger than about 4 eV, the solid is an insulator. Most semiconductors have band gap energies from some tenths of an eV
to about 3.5 eV. For comparison, the thermal energy kT at
room temperature amounts to 0.026 eV.

In semiconductors, both electrons in the conduction band (the
lowest empty band) and defect electrons, called holes, in the
valence band (the highest occupied band) can contribute to
electrical conduction. If electrons from the valence band are
excited thermally to the conduction band, the intrinsic conduction with equal numbers of electrons and holes results.
This type of conduction is, however, only found with extremely pure semiconductors and/or at very high temperatures. Usually extrinsic n- or p-type conduction is predominant, which
is caused by impurity atoms. Donor atoms give rise to energy
levels close to the conduction band, so that they are fully
ionized at ordinary temperatures, freeing one of their outer
electrons into the conduction band. Similarly, acceptor levels
are located close to the valence band, giving rise to hole
conduction.

Considering the energy bands in more detail, the dependence
of the electron or hole energy on the wavevector k, which is
proportional to the momentum, has to be taken into account.
Two energy wavevector diagrams are shown in Fig. 1-5. With
the example on the left hand side, both bands exhibit parabolic shapes. The minimum of the conduction band and the maximum of valence band coincide at zero wavevector. A material
with this type of band diagram is called a direct semiconductor. In contrast to this behaviour, an indirect semiconductor
is characterized by a band diagram similar to the right hand
side of Fig. 1-5. Here the conduction band exhibits two mini-

ma, and the lowest minimum, which determines the band gap, does not occur at zero wavevector. The transition with the smallest change in energy causes a simultaneous change in wavevector. This is called an indirect transition, whereas a direct transition takes place at constant wavevector. Examples of direct and indirect semiconductors will be given in chapter 2.

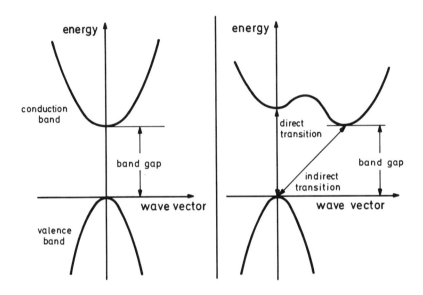

Fig. 1-5     Energy wavevector diagrams for a direct
             and an indirect semiconductor, showing
             direct and indirect transitions

## 1-3 Absorption and recombination

This section deals with the basic interaction processes of light with semiconductors, absorption and recombination. Photons with energies smaller than the band gap travel through a pure semiconductor almost without interaction: the semiconductor is transparent for light with corresponding wave-

lengths. Photons with energies larger than $E_g$ can lift an
electron from the valence to the conduction band, thus generating an electron hole pair. This absorption process can be
written like a chemical reaction:

$$photon \longrightarrow electron \ + \ hole$$

In direct semiconductors this process can take place exactly
as just described. In indirect materials, however, the electron hole pair generation requires a certain momentum in addition to the band gap energy. Because the photon has almost
zero momentum, an additional particle must be either absorbed
or generated:

$$photon \ +/- \ phonon \longrightarrow electron \ + \ hole$$

Phonons are quantum particles related to lattice vibrations;
the energy of acoustical phonons is of the order of the thermal energy kT, which is in most cases much smaller than the
photon energy. Because the absorption process in indirect
semiconductor materials requires an additional particle, it
is a higher order process with lower probability than the
simpler process in direct semiconductors. This difference can
be seen clearly from Fig. 1-6, where the absorption coefficients of two important semiconductors are plotted as functions of the photon energy. The absorption coefficient $\alpha$ is
defined by the exponential factor $\exp(-\alpha t)$, which describes
the attenuation of light after travelling through a sample of
thickness t. GaAs is a direct semiconductor with a band gap
of 1.42 eV. As shown in Fig. 1-6, the absorption coefficient
rises very steeply around $E_g$, reaching values above $10^4$ cm$^{-1}$
for energies larger than $E_g$. In contrast to the steep absorption edge of a direct semiconductor, the absorption coefficient of GaP, an indirect material with $E_g$ = 2.26 eV, rises
much more gradually, showing a shoulder above $10^2$ cm$^{-1}$. Only

at energies around 2.8 eV can a further rise be seen due to
the onset of direct absorption. The reciprocal of the absorption coefficient is the absorption length, the light path at-
tenuating the light by a factor of 1/e. The absorption length
in direct semiconductors can be very short: $10^{-4}$ cm = 1 μm
for hv $\geq$ $E_g$. In indirect materials much larger values are typical: $\alpha$ = $10^{-2}$ cm$^{-1}$ at hv $\approx$ $E_g$, giving an absorption length
of 100 μm.

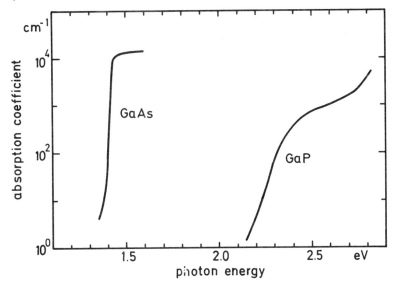

Fig. 1-6    Absorption coefficients of GaAs (direct)
            and GaP (indirect) as functions of photon
            energy

Compared with absorption, recombination is an opposite process: an electron and a hole recombine with emission of a
photon. In direct semiconductors, radiative recombination
occurs with high efficiency:

        electron  +  hole $\longrightarrow$ photon

In indirect semiconductors, energy and momentum conservation
again requires the participation of a phonon:

electron  +  hole ⟶ photon  +/- phonon

The recombination probability is conventionally described
using the minority carrier lifetime $\tau$. The radiative lifetime
in direct materials is of the order of $10^{-9}$ s, corresponding
to a high radiative efficiency. Radiative lifetimes in indirect semiconductors can be much larger, for example $10^{-6}$ s.
Because this lifetime is considerably larger, competing nonradiative recombination processes can take place. If radiative recombination (characterized by $\tau_r$) and nonradiative recombination ($\tau_{nr}$) occur simultaneously, the effective carrier
lifetime $\tau$ is given by:

$$1/\tau = 1/\tau_r + 1/\tau_{nr} \qquad\qquad (1-1)$$

as derived in section 6-1-2. From this type of equation it
follows that the shorter value of $\tau_r$ or $\tau_{nr}$ determines the
effective lifetime. In LEDs the radiative efficiency is frequently much smaller than 100 %. That means that $\tau_r$ is much
larger than $\tau_{nr}$, and $\tau$ is approximately $\tau_{nr}$.

In Fig. 1-7 three examples of recombination processes in connection with the band diagram are shown. The radiative direct
transition is described by the vertical arrow on the left
hand side: the electron drops from the conduction to the valence band. By comparison, a typical nonradiative transition
(broken arrow) takes place via a deep state due to an impurity atom. In this case the recombination energy is given to a
multitude of phonons:

electron  +  hole ⟶ n phonons

The third example, recombination via a localized shallow
state, is important in indirect semiconductors. The electron
first drops into the shallow state, losing only a small part

of its energy. Because the electron is localized there, its wave function is spread out in k space. This is due to the well-known Heisenberg uncertainty principle which states that the position and the momentum of a particle cannot be determined exactly at the same time. Because the wave function is spread out in k space, it also has a noticeable contribution at k = 0. Therefore, the localized electron can recombine radiatively with a hole in the valence band.

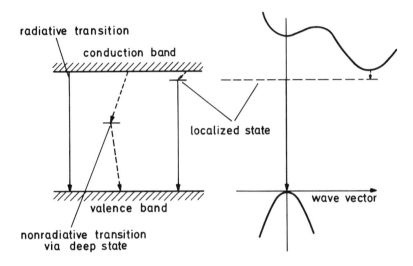

Fig. 1-7     Radiative and nonradiative recombination processes in direct and indirect semiconductors

## 1-4 pn junction

In this section the most important properties of semiconductor pn junctions are briefly reviewed.

In p material the Fermi level is close to the valence band, and in n material close to the conduction band. Because in thermal equilibrium the Fermi level must be constant within a semiconductor, the bands at a pn junction without external bias must be bent as shown in Fig. 1-8, upper part. The region around the pn junction is depleted of mobile carriers, the ionized donors and acceptors form space charges here (space charge region), whereas the p and n regions are neutral. If an external bias is applied in reverse direction (p region negative, n region positive), the barrier height is enlarged, and the space charge region becomes wider. With forward bias (p region positive, n region negative) the barrier height is reduced, and the space charge region is narrowed. If the forward bias is comparable with $E_g/e$, the barrier height becomes so small that large numbers of electrons are injected into the p region, and holes into the n region (see Fig. 1-8, lower part). The injected minority carriers recombine with the majority carriers. The spatial decay of the minority carrier densities is described by an exponential factor $\exp(-x/L)$, with distance x and characteristic

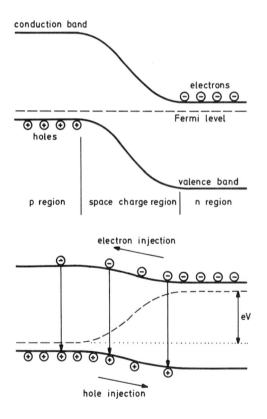

Fig. 1-8 Energy bands at a pn junction in thermal equilibrium
(top) and in a forward biased state (bottom)

length L (diffusion length). L is related to the minority
carrier lifetime $\tau$:

$$L = \sqrt{D\tau}  \qquad (1\text{-}2)$$

with the diffusion constant $D = \frac{kT}{e} \mu$, $\mu$ = carrier mobility.

In most cases the current flowing accross a pn junction practically consists of only one type of carrier. Which carrier
type dominates depends normally on the doping levels ( in a
$p^+n$-diode hole injection is preferred), and the carrier mobilities (electron injection in the case of $\mu_e \gg \mu_h$, e.g. in
GaAs). In the case of heterojunctions, however, where p and n
regions have different band gaps, carrier injection takes
place from the higher to the lower band gap region. Injection
in the opposite direction is prevented by an additional barrier of height $\Delta E_g$, the difference of band gaps, which decreases current flow by a factor of exp $(\Delta E_g/kT)$.

The I V characteristics of a normal pn diode for small voltages is described by the equation

$$I = I_0 \left\{ \exp(eV/nkT) - 1 \right\}  \qquad (1\text{-}3)$$

with $I_0$ = saturation current and n = numeric factor, which
equals 1 for normal injection and 2 in the case of recombination in the space charge region or at surfaces. In practice,
n is somewhere between 1 and 2. These characteristics are
shown in Fig. 1-9. For negative voltages larger than several
kT/e the current is practically constant at $-I_0$, and for positive voltages larger than some kT/e the current rises exponentially, i.e. $I \sim \exp(eV/nkT)$. Deviations from this ideal
behaviour are found in all real diodes. Two important examples are included in Fig. 1-9: first, reverse breakdown occurs at some reverse voltage (mainly avalanche breakdown for

lightly doped semiconductors), and second, the exponential rise of the current is diminished at high currents, due to a series resistance.

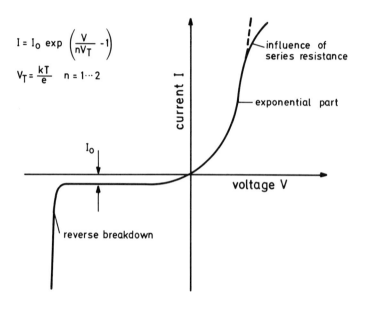

Fig. 1-9    I-V characteristics of a pn diode, also showing deviations from ideal behaviour

Other common deviations from ideal characteristics are leakage currents, which are especially noticeable in the reverse direction, and the onset of high injection with n approaching 2 at high currents.


## Questions

Q1-1        What is the number of nearest neighbours in the diamond lattice?

Q1-2      What percentage of the volume of a Si crystal is
          filled with atoms, assuming the atoms to be spheres
          touching each other?

Q1-3      What is the number of Ga or As atoms per $cm^3$ in a
          GaAs crystal (zincblende type, lattice constant
          0.5653 nm)?

Q1-4      Sulfur is a donor impurity in GaAs, which occupies
          As sites. What percentage of the As atoms has to be
          replaced by S atoms, to achieve an n-type doping
          with $10^{18}$ electrons per $cm^3$, assuming every S atom
          to give one electron into the conduction band?

Q1-5      What is the effective carrier lifetime in a semi-
          conductor if the radiative recombination is char-
          acterized by a radiative lifetime $\tau_r$ = 1 μs, and
          the nonradiative recombination by $\tau_{nr}$ = 50 ns?

Q1-6      What is the diffusion length of injected electrons
          in p GaAs at room temperature if the mobility of
          electrons in p material is assumed to be 4000
          $cm^2$/Vs, and if the lifetime is 50 ns?

# 2 MATERIALS FOR LEDS

In this chapter we will describe the requirements which have to be fulfilled by materials for light emitting diodes, and the relevant properties of several semiconducting compounds. Emphasis is laid upon the materials which are in use in the modern semiconductor industry, however, more exotic materials will also be treated briefly.

## 2-1 General requirements

The requirements to be met by LED materials can be divided into three groups:

(1) absolute requirements which have to be fulfilled in all cases because they are based on physical laws,

(2) practical requirements which are not of a principle nature, but are in practice met by all materials for light emitting diodes,

(3) economic requirements which necessitate the fabrication of LEDs by as simple means as possible, i. e. at low cost. This is of course especially important for large volume production.

The most obvious requirement from group (1) is that the band gap energy of the material must be at least equal to the required photon energy. If the emitted radiation is characterized by the wavelength $\lambda$, the minimum band gap energy $E_g$ is:

$$E_g = \frac{hc}{\lambda} \text{ , or } E_g \text{ (in eV)} = \frac{1240}{\lambda \text{ (in nm)}} \qquad (2\text{-}1)$$

because $\frac{hc}{e} = 1.24 \cdot 10^{-6}$ V m $= 1240$ V nm. For visible LEDs with wavelengths between 400 and 700 nm, semiconductors with band gap energies between 3.1 and 1.77 eV respectively are therefore required. The common elemental semiconductors germanium ($E_g = 0.67$ eV) and silicon ($E_g = 1.15$ eV) have band gap energies which are too small for this purpose.

A second absolute requirement is that an efficient radiative recombination mechanism must exist in LED materials. This can be best accomplished by direct band-to-band transitions, i.e. direct semiconductors are preferred. Germanium and silicon are both indirect materials. Recombination via a localized shallow state (see section 1-3) is an inferior but nevertheless practicable mechanism for light generation in some indirect materials (see section 2-2). Still less efficient in indirect semiconductors is recombination involving donors and/or acceptors. Third, it must be possible to dope the material to exhibit n-type and p-type conduction. Otherwise, the required pn junctions cannot be fabricated.

The most important requirement from group (2) is that the material can be produced in single crystalline form, either as bulk crystal or as epitaxial layer. All existing light emitting diodes are based on single crystalline materials although there is no reason in principle why LEDs could not be fabricated from polycrystalline or even amorphous semiconductors. In the case of epitaxial material, an appropriate substrate material must be available.

The economic requirements, group (3), are numerous, and only some examples will be mentioned here: it must be possible to produce bulk or substrate crystals of sufficient size to allow economic production of LEDs. Epitaxial growth methods with reasonable process conditions and growth rates should exist, and finally, technological processes like diffusion doping, etching, and ohmic contact fabrication should be developed to the stage where they can be applied without major difficulties.

## 2-2 Binary III-V compounds

Because light emitting diodes cannot be fabricated from germanium or silicon, other semiconductors have to be used. The basic materials for nearly all LEDs belong to the large class of III-V compounds, which were investigated first by WELKER (1952). The III-V compounds can be considered to be derived from the elemental semiconductors in the following manner (see Fig. 2-1): germanium, an element from group IV of the periodic table of the elements, has 32 electrons, four of which form tetrahedally oriented covalent bonds to the neighbouring atoms, resulting in a diamond lattice type crystal (see section 1-1). If every second germanium atom is replaced by its left hand neighbour in the periodic table, which is gallium with three valence electrons, and the other half of germanium atoms is replaced by arsenic atoms, the opposite neighbour with five valence electrons, the resulting crystal consists of the compound gallium arsenide, GaAs. The average number of valence electrons per atom is again four, so that GaAs should have similar properties to Ge. Indeed, GaAs is a semiconductor; however, it has a direct band gap which is larger than that of Ge. In a similar way, aluminum phosphide

AlP is related to silicon.

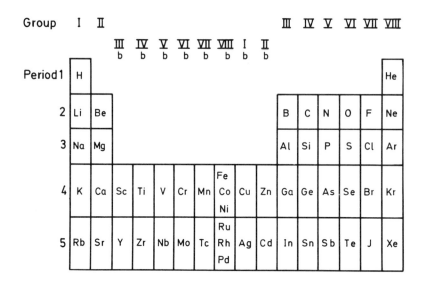

Fig. 2-1      Periodic table of the elements, periods 1
              to 5

Generally, by combination of the group III elements Al, Ga,
In with the group V elements N, P, As, Sb, twelve binary
III-V compounds can be formed as shown in Fig. 2-2. The number directly below the chemical formula in Fig. 2-2 gives the
band gap energy in eV, with direct gaps marked with an aster-
isk. The band gaps increase from right to left, and from bottom to top, i.e. with decreasing atomic numbers of both constituents, parallel to the increase in binding energy. With
the compounds from InSb (0.18 eV) to GaN (3.4 eV) the wavelength range from 6.89 μm to 365 nm can in principle be covered. AlN with a bandgap of approximately 6 eV is an insulator and cannot be considered for the fabrication of LEDs.

Direct gaps are found with the smaller band gaps, and GaAs
has the highest direct gap, not considering the nitrides,

| | N | P | As | Sb | |
|---|---|---|---|---|---|
| Al | Al N<br>~ 6.0 ?<br>(0.440) | Al P<br>2.45<br>0.546 | Al As<br>2.15<br>0.566 | Al Sb<br>1.65<br>0.614 | eV<br>nm |
| Ga | Ga N<br>3.4*<br>(0.451) | Ga P<br>2.26<br>0.545 | Ga As<br>1.42*<br>0.565 | Ga Sb<br>0.73*<br>0.610 | eV<br>nm |
| In | In N<br>1.95*<br>(0.501) | In P<br>1.34*<br>0.587 | In As<br>0.36*<br>0.606 | In Sb<br>0.18*<br>0.648 | eV<br>nm |

Fig. 2-2    Band gap energies and lattice constants of
            binary III-V compounds

which are also exceptions with respect to crystal structure.
The III-V compounds (without nitrides) crystallize in the
cubic zincblende lattice (see section 1-1); the lattice constants in nm as displayed in fig. 2-2 show an opposite tendency to the band gap energies. The nitrides crystallize in
the hexagonal wurtzite lattice, and the numbers shown in
brackets are the hypothetical cubic lattice constants assuming a zincblende lattice with the same nearest neighbour
distance.

Although GaP is an indirect semiconductor, the radiative recombination efficiency in this material can be greatly enhanced by introduction of isoelectronic centers. If GaP is
doped with nitrogen, some of the phosphorus atoms are replaced with nitrogen. Because nitrogen has the same number of va-

lence electrons as phosphorus, it does not contribute to
electronic conduction. However, it introduces a shallow level
below the conduction band which can localize an electron,
which in turn attracts a hole, forming a bound exciton. This
bound exciton recombines radiatively, giving rise to emission
of green light with photon energies some 50 meV below the
band gap. The increase in efficiency by nitrogen doping is
due to the localization of the electron (see section 1-3).
Green LEDs on the basis of GaP rely on this mechanism (see
section 4-1-3). By simultaneous doping of GaP with zinc and
oxygen, another isoelectronic center can be introduced. It
consists of one atom each of Zn and O on adjacent lattice
sites and emits red light (1.8 eV, 690 nm) with high efficiency. One type of red LED uses this mechanism for light
generation (see section 4-1-1). Only these two examples of
isoelectronic centers are of practical importance. With other
materials and other impurities the gain in efficiency is much
lower.

The binary III-V compounds with zincblende structure can be
doped n or p type. The group VI elements sulfur, selenium and
tellurium (see Fig. 2-1) can replace the group V constituent,
e.g. As in GaAs. Because they introduce an additional valence
electron, n-type conduction results. The donor energies are
usually below 100 meV, so that they are fully ionized at room
temperature. Oxygen forms a deep donor which is not useful
for device purposes. Similarly, the group II elements beryl-
lium, magnesium, zinc, and cadmium form shallow acceptors,
because they contribute one electron less than the group III
constituent which they replace. The group IV elements carbon,
silicon, germanium, and tin can be incorporated on group III
or group V sites, where they act as donors or acceptors, respectively. The conditions predominating during incorporation
(temperature, pressure, etc.), determine which mechanism dominates, however, Sn is more likely to be a donor, and C is

preferentially an acceptor. The doping behaviour of AlN and InN is hardly known; up to the present time GaN can only be made n type or insulating, probably due to an autocompensation mechanism. The problems related to GaN are treated in section 4-1-4.

As mentioned in section 2-1, single crystalline material is required for fabrication of LEDs. Bulk crystals of GaAs, GaP, and InP are grown on an industrial scale and are easily available with different doping. Crystals of InAs, InSb and GaSb are also supplied by some manufacturers. AlP, AlAs and AlSb could be grown as single crystals; however, these compounds decompose on contact with moisture. Therefore, they cannot be handled easily and are excluded for practical reasons. Only very small crystals of the nitrides AlN, GaN and InN have been grown on a laboratory scale.


## 2-3 Mixed crystals

Because the III-V compounds are rather similar to each other, they can form mixed crystals with properties between the binary compounds. This is an extremely useful property because it allows the properties to be tailored for a specific purpose. We will present here two examples of the ternary compounds and one of the quaternary compounds.

If x moles (x is a number smaller than 1) of e.g. GaP are mixed with (1-x) moles of GaAs, the resulting mixed crystal is described by the formula $GaAs_{1-x}P_x$. The parameter x gives the mole fraction of P, x = 0 corresponds to pure GaAs and x = 1 to GaP. Two important properties of $GaAs_{1-x}P_x$ as a function of the composition x are plotted in Fig. 2-3. The band gap changes continuously between x = 0 and x = 1, it is direct up to x = 0.45 ($E_g$ = 1.99 eV) and indirect in the

range 0.45 < x < 1. The kink at x = 0.45 is due to the cross-
over of two different band gaps. Similarly to GaP (see section 2-2), GaAs$_{1-x}$P$_x$ in the indirect range can be doped with
the iso-electronic impurity nitrogen, so that radiative recombination with reasonable efficiency is possible for all
compositions of GaAs$_{1-x}$P$_x$. The lattice constant varies line-
arly between 0.565 nm (GaAs) and 0.545 nm (GaP). This linear
dependence of the lattice constant on composition is known as
VEGARD's law. Because of the lattice constant variation, epitaxial layers of GaAs$_{1-x}$P$_x$ can be grown on GaAs or GaP substrates only by using special measures (see section 3-2-2).

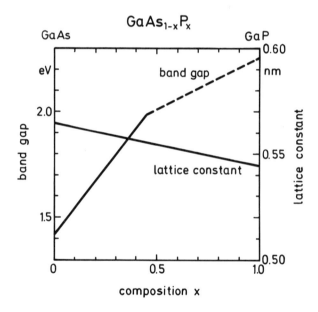

Fig. 2-3      Band gap energy and lattice constant of
              gallium arsenide phosphide

A second example of a ternary compound is shown in Fig. 2-4,
which describes properties of gallium aluminum arsenide
Ga$_{1-x}$Al$_x$As. The composition dependence of the band gap looks
very similar to the last example. The direct-indirect transition occurs at x = 0.44, E$_g$ = 1.96 eV. Because no efficiency-

enhancing isoelectronic impurity is known in this system,
only the direct gap range $0 \le x \le 0.44$ can be used for light
emitting diodes. An important peculiarity of $Ga_{1-x}Al_xAs$ is
that the lattice constant is nearly independent of composition. This property eases epitaxial growth, (see section
3-2-1) because any composition of $Ga_{1-x}Al_xAs$ can be grown
directly on to GaAs substrates and on any other composition
of $Ga_{1-x}Al_xAs$, which is important for the fabrication of
heterostructures.

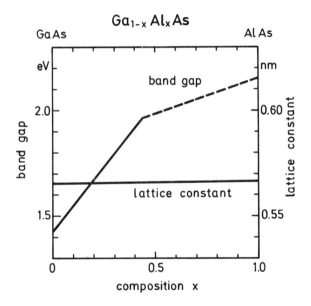

Fig. 2-4    Band gap energy and lattice constant of
            gallium aluminum arsenide

Another mixed crystal system, which is used for infrared
emitting diodes and lasers, is the quaternary
$In_{1-x}Ga_xAs_{1-y}P_y$, the band gap variation of which is shown in
Fig. 2-5. Here the composition is described by two parameters: x gives the composition of the group III constituents,
and y of the group V constituents. Because the range of compositions is two-dimensional, the band gap has to be plotted

in the third dimension. In Fig. 2-5 iso-band gap lines are
shown for the direct region. The lowest band gap is in the
InAs corner, and the highest direct band gap occurs at the
InP-GaP boundary: 2.2 eV at $In_{0.73}Ga_{0.27}P$. The second important property of $In_{1-x}Ga_x As_{1-y}P_y$, the lattice constant,
could be plotted with another set of lines in the diagram;
however, only one iso-lattice constant line is included in
Fig. 2-5 (broken line). This line marks the compositions with
exactly the same lattice constant as InP. As it can be seen
easily, compositions with band energies from 0.7 eV to
1.35 eV can be lattice-matched to InP; the corresponding
wavelength range is 1.77 μm to 919 nm.

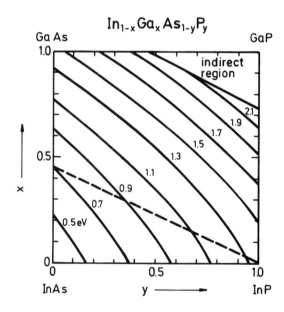

Fig. 2-5     Iso-band gap lines of indium gallium ar-
             senide phosphide

Compared with ternary systems, quaternary systems offer an
additional degree of freedom in the choice of materials. Because it is difficult to control the composition of mixed
crystals with high accuracy over a large volume of material,

bulk crystals have only been grown in research laboratories
and are not readily available. Instead, ternary and quaternary materials are produced by epitaxial techniques (section
3-2).


## 2-4 Other materials

With the binary and ternary III-V compounds described in
sections 2-2 and 2-3 emitters with wavelengths from the infrared to the green part of visible spectrum have been realized and are produced in large numbers.

However, other materials have also been investigated with one
of the following two aims:

(1)   to find materials with improved luminescence effi-
      ciency in the standard wavelength range,

(2)   to develop materials for blue-green to blue light
      emitting diodes.

In this chapter we want to give some examples for such in-
vestigations and describe briefly the present status.

### $In_{1-x}Ga_xP$

This compound has a considerably larger direct-indirect transition than $GaAs_{1-x}P_x$. The maximum direct gap is 2.2 eV at
x = 0.73, so that efficient red to yellow LEDs should be possible. The lattice constant is however strongly composition
dependent, and the enthalpies of formation of the binaries
are very different, making epitaxial growth of this compound
difficult. $In_{0.48}Ga_{0.52}P$ with a bandgap of 1.97 eV is lattice
matched to GaAs, for other compositions, however, substrate

crystals are not available. Due to these difficulties the
theoretical advantages over $GaAs_{1-x}P_x$ have not been realized
up to now.

## $In_{1-x}Al_xP$ and $In_{1-x-y}Ga_xAl_yP$

$In_{1-x}Al_xP$ has the largest direct-indirect transition of all
III-V compounds: 2.33 eV at x = 0.44, corresponding to green
emission. With addition of gallium the quaternary composition
$In_{0.45}Ga_{0.34}Al_{0.21}P$ still has a direct gap of 2.17 eV and the
same lattice constant as GaAs. Using new epitaxial techniques
(see section 3-2-3) the extreme thermodynamical problems,
which have inhibited the practical use of this compound,
could possibly be circumvented.

## $Ga_{1-x}Al_xP$

This compound is indirect for all compositions, because both
GaP and AlP have indirect gaps, but its band gap is larger
than that of GaP. Therefore, it was considered as a material
for deep green LEDs. Attempts to dope it with the isoelectronic center nitrogen were however unsuccessful due to AlN formation.

For blue LEDs materials with much larger band gap energies
are required. One candidate, GaN, has already been treated in
section 2-2. Further possibilities are silicon carbide, SiC,
and zinc sulfide selenide, $ZnS_xSe_{1-x}$.

## SiC

Silicon carbide is a IV-IV compound which exhibits many
slightly different crystallographic forms (polytypism). The
indirect band gap energies lie between 2.3 and 3.3 eV, and

one of the most frequent polytypes, the hexagonal 6H, has a band gap of 2.8 eV. Because the band is indirect and luminescence enhancing isoelectronic centers are not known, recombination takes place with lower radiative efficiency involving donors and/or acceptors. Nevertheless, blue LEDs on the basis of SiC are available (section 4-1-4). The technology of this material involves very high temperatures around 2000° C and, in particular, crystal growth of SiC is very difficult, yielding only platelets with dimensions of typically 1 cm maximum. In addition, the material is extremely hard and can only be machined with diamond tools.

## $ZnS_xSe_{1-x}$

Zinc sulfide selenide belongs to the class of II-VI compounds which have direct and generally larger band gaps than the III-Vs. These materials show a strong tendency for self-compensation, so that they can be produced with only one type of conductivity. CdTe which can be n or p type is the only exception to this rule; ZnTe is always p type, and all other compounds can only be made with n type conductivity. This is also true for ZnS and ZnSe. At one time it was claimed that pn junctions could be generated in $ZnS_xSe_{1-x}$ produced by a special vapor phase transport reaction; these results were however not confirmed. Bulk crystals of the II-VI compounds can be produced by sublimation methods, but are not commercially available. Epitaxial layers of II-VI compounds can also be grown on substrate crystals of standard semiconductors, e.g. GaP. To summarize, LEDs on the basis of II-VI compounds are still far from practical realization.

## Questions

Q2-1    What is the emission wavelength of a hypothetical LED
        made from silicon?

Q2-2    Why are direct semiconductor materials preferred for
        LED fabrication?

Q2-3    Which III-V compounds can be considered to have the
        largest degree of similarity to diamond and tin, re-
        spectively?

Q2-4    What is the advantage of using mixed crystals?

Q2-5    Why is the lattice constant of an LED material so im-
        portant?

Q2-6    What is the lattice constant of $GaAs_{0.6}P_{0.4}$?

Q2-7    Which composition of $In_{1-x}Ga_xAs_{1-y}P_y$ emits infrared
        radiation of 1.378 µm wavelength and has the same lat-
        tice constant as InP?

Q2-8    III-V and II-VI compounds have been mentioned in this
        chapter. Which class of compounds could also be ex-
        pected to be semiconductors, but are not, and why
        aren't they semiconducting?

## References

ARCHER, R.J., "Materials for light emitting diodes,"
    Journal of Electronic Materials, 1 (1972),
    pp. 1-26

BERGH, A.A., and P.J. DEAN, "Light emitting diodes,"
    Proceedings of the IEEE, 60 (1972), pp. 156-223

CASEY, H.C., and F.A. TRUMBORE, "Single crystal electrolumi-
    nescent materials," Materials Science and Engineering,
    6 (1970), pp. 69-109

NUESE, C.J., "III-V alloys for optoelectronic applications,"
    Journal of Electronic Materials, 6 (1977),
    pp. 253-293

WELKER, H., "Über neue halbleitende Verbindungen,"
    Zeitschrift für Naturforschung, 7a (1952),
    pp. 744-749

# 3 LED TECHNOLOGY

This chapter deals with the techniques used for fabrication of LEDs, with emphasis on the most commonly used methods. It is hardly possible and also not necessary to describe all variations of LED technology within the frame of this book; we will therefore concentrate on the methods which are in our view most important. For example, the technology of blue LEDs is mentioned briefly elsewhere in this book (sections 2-4 and 4-1-4) and is excluded here. Our aim is mainly to present an overview of LED technology, and to explain why a particular process is being used for a particular fabrication step.

In Fig. 3-1 a brief summary of the technology of GaAs and GaP based LEDs is shown. The first step is the growth of single crystals to be used as substrates. There are two growth methods available: the horizontal Bridgman (HB) or boat growth, and the liquid encapsulation Czochralski (LEC) method, which is a kind of crucible pulling method. GaAs crystals can be grown by both methods (HB being preferred), whereas GaP is grown exclusively by LEC. Because the quality of these bulk crystals is inadequate for LEDs, and because ternary materials are difficult to grow in bulk form, the active material for LEDs is produced by an epitaxial method, either liquid phase epitaxy (LPE) or vapor phase epitaxy (VPE). LPE is used with binary compounds (GaAs and GaP) and with $Ga_{1-x}Al_xAs$, i.e. in all cases where lattice matching is no problem. VPE has to be applied for $GaAs_{1-x}P_x$, which is deposited on GaAs

| crystal growth | boat (HB) | | | crucible (LEC) | | | |
|---|---|---|---|---|---|---|---|
| substrate material | GaAs | | | GaP | | | |
| epitaxy | LPE | | VPE | | | LPE | |
| doping n<br>  " p | Si<br>Si | Te<br>Zn | Te<br>– | S<br>– | | S<br>Zn | Te<br>Zn |
| active material | GaAs | (GaAl)As<br>8% Al | (GaAl)As<br>40% Al | Ga(AsP)<br>40% P | Ga(AsP):N<br>60% P | Ga(AsP):N<br>85% P | GaP:N | GaP:Zn,O |
| diffusion | – | | Zn | | | – | |
| wavelength in nm | 950 | 870 | 650 | 660 | 625 | 590 | 565 | 690 |
| color | – | – | red | red | orange | yellow | green | red |

Fig. 3-1     Overview of LED technology

substrates for x < 0.5 and on GaP substrates for x > 0.5,
because minimum lattice mismatch is achieved in this way.
Liquid phase epitaxial layers usually contain a grown-in pn
junction, whereas VPE material is normally grown n type, and
the pn junction is produced by zinc diffusion later. With
these materials and technologies the wavelength range from
950 nm (IR) to 565 nm (green) can be covered.


## 3-1 Synthesis and bulk crystal growth

There are three binary III-V compounds which are used as substrates in LED fabrication: GaAs, GaP and to a smaller extent
InP. The four chemical elements Ga, In, As, P necessary to
synthesize these compounds are found in many natural miner-
als, and we want to mention the most important sources. Gallium is a relative of aluminum; it is therefore contained in

aluminum ores, e.g. bauxite. In fact, gallium is a by-product
of aluminum extraction. Indium occurs mainly as an admixture
to zincblende (ZnS); it is consequently produced in conjunction with the smelting of zinc. The most important sources of
arsenic and phosphorus are arsenopyrite FeAsS and apatite
$Ca_5F(PO_4)_3$. The raw elements have to be purified to an extremely high degree before they can be used for semiconductor
purposes. The first cleaning steps are mostly suitable chemical reactions; the final purity is however achieved by physical methods such as distillation, sublimation and recrystallization. For example, high purity gallium is produced by repeated recrystallization, a process which is favoured by the
low melting point of this element (about 30 °C).

The technical difficulties in crystal growth of a particular
III-V-compound are mainly determined by its melting point and
the dissociation pressure at the melting point. These properties and some other relevant data of LED substrate materials
are summarized in Table 3-1.

Table 3-1

Properties of substrate materials

|  |  | GaAs | GaP | InP |
|---|---|---|---|---|
| melting point | (°C) | 1238 | 1470 | 1062 |
| dissociation pressure | (bar) | 0.98 | 35 | 27 |
| lattice constant | (nm) | 0.565 | 0.545 | 0.587 |
| molecular weight |  | 144.6 | 100.7 | 145.8 |
| density | $(g/cm^3)$ | 5.32 | 4.13 | 4.79 |
| thermal expansion coeff. | $(10^{-6}K^{-1})$ | 5.9 | 4.7 | 4.5 |

The melting points of GaAs, GaP and InP allow crystal growth
from the melts in a similar way to the elemental semiconduc-

tors Si and Ge; however, special measures against decomposition are necessary with the III-V compounds. Without these
measures the high dissociation pressures would cause complete
volatilization of the group V component.

For growth of gallium arsenide the horizontal Bridgman (HB)
method is widely used, and this is depicted in Fig. 3-2.

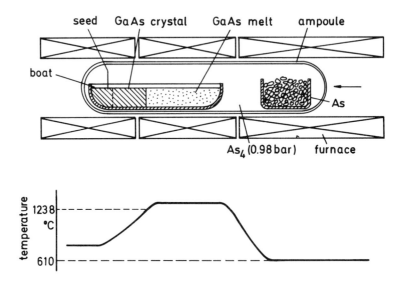

Fig. 3-2     Horizontal Bridgman growth of GaAs

It is a combined synthesis and crystal growth method taking
place in a boat which consists usually of quartz. The boat
containing gallium metal at the beginning of the process is
placed in a closed quartz ampoule together with a reservoir
of arsenic which is kept at a temperature of 610 °C, which
results in an arsenic vapor pressure of 0.98 bar. The gallium
is heated to a temperature slightly above the melting point
of GaAs. It reacts with the arsenic vapor until the whole
metal is transformed into GaAs. The amount of As is chosen so
that some As is left after complete reaction to keep a con-

stant arsenic pressure during the following crystal growth
process. This is initiated by bringing the GaAs melt into
contact with a GaAs seed crystal (for example by slightly
tilting the ampoule) and by slow movement of the ampoule
through a temperature gradient. Alternatively, the movement
of the ampoule can be simulated by corresponding change of
the temperature profile within a multi-zone furnace. By proper control of the crystal-melt interface shape and of the
crystallization speed the complete content of the boat
freezes as a single crystal with the orientation determined
by the seed. Crystals with masses of several kg and diameters
from 50 to 80 mm are routinely grown by many suppliers.

The HB process yields GaAs crystals of high quality. Dislocation densities of this material are usually in the range $10^3$
to $10^4$ $cm^{-2}$, which means that a typical LED with an area of
$10^{-3}$ $cm^2$ contains only 1 to 10 dislocations. The shape of HB
crystals is determined by the cross section of the boat.
Therefore, GaAs slices cut from a boat grown crystal mostly
exhibit a half-circular "D"-shape.

For mass production round slices are advantageous. They can
be produced using the LEC process which is explained below
using GaP as an example. This method yields crystals with
circular cross section and diameters in the same range, the
dislocation densities are however considerably higher, namely
$2 \times 10^4$ to $5 \times 10^4$ $cm^{-2}$ or even higher for crystal diameters
approaching 80 mm.

It is impossible to grow gallium phosphide crystals using the
horizontal Bridgman method because quartz parts cannot with-
stand temperatures close to 1500 °C and pressures of 35 bar
simultaneously. Instead, GaP is synthesized in a separate
process at somewhat lower temperature and pressure (about
1300 °C, 10 bar). From the polycrystalline starting material

GaP single crystals are grown using the liquid encapsulation
Czochralski (LEC) method which is shown in Fig. 3-3. Here the
melt is contained in a crucible which is heated inductively
or with a resistance heater and is placed in a pressure vessel which is filled with an inert gas (e.g. nitrogen) at a
pressure higher than the dissociation pressure of GaP. To
suppress vaporization of phosphorus, the melt is covered with
a layer of molten boron oxide $B_2O_3$, a compound with a unique
combination of properties which are necessary for this purpose: it has a rather low melting point and vapor pressure,
it is lighter than GaP and transparent and is inert against
the other materials involved. Growth of a single crystal is
accomplished by dipping a seed crystal through the encapsulation layer and slowly
withdrawing it with a
typical speed of 0.5 mm
per minute. The growing
crystal is usually ro-
tated to equalize tempe-
rature differences. For
safety reasons the pro-
cess is monitored using a
TV system. GaP crystals
grown by the LEC process
are available with dia-
meters of about 50 mm and
dislocation densities
around $10^5$ cm$^{-2}$.

Indium phosphide crystals
are produced similarly to
GaP. An additional diffi-
culty arises from the

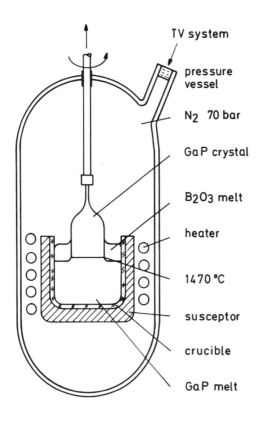

TV system

pressure
vessel

N$_2$ 70 bar

GaP crystal

B$_2$O$_3$ melt

heater

1470 °C

susceptor

crucible

GaP melt

Fig. 3-3   Liquid encapsulation Czochralski growth of GaP

rather low melting point of InP. At this temperature the viscosity of the $B_2O_3$ encapsulant is considerably higher than at
the melting point of GaP. Together with a lower stacking
fault energy, this results in a higher tendency for twin formation during crystal growth, diminishing the yield of usable
single crystals.

If GaAs is grown by the LEC technique, the separate synthesis
process can be avoided by application of the so-called direct
synthesis method. Here the elements gallium and arsenic are
loaded in the crucible together with the boron oxide. On
heating under high pressure (e.g. 100 bar) the $B_2O_3$ melts and
seals the other materials before noticeable vaporization of
As takes place. This approach is mainly used for production
of semi-insulating GaAs for electronic applications.

Growth from a nonstoichiometric melt at temperatures considerably below the melting point and consequently much lower
pressures was investigated as an alternative for crystal
growth of the phosphides GaP and InP. This technique is known
as temperature gradient solution (TGS) or synthesis, solute
diffusion (SSD) method. Single crystalline growth is more
difficult to achieve with this method, however, and the
growth rate is much smaller than with LEC growth because
phosphorus has to be transported by diffusion through the
melt to the growing interface. Therefore this technique did
not prevail.

For use as LED substrates the III-V crystals have to be doped
n type in most cases, only some LED-types require p type substrates. N type doping is achieved by addition of appropriate
amounts of silicon, sulfur, tellurium or compounds of these
elements to the melt, whereas zinc is used for p type doping.
The incorporation of these impurities during crystal growth
is not constant, however, due to segregation effects. As a

result, the doping level varies within a crystal boule from one end to the other. This is not a serious restriction of the applicability of III-V crystals, however, because the LED properties do not depend critically on the substrate doping.

For the following process steps the substrate crystals are cut with diamond saws into slices of typically 350 μm thickness with (111) or (100) orientation depending on the type of LED to be produced. The surfaces of the slices are usually polished and etched to remove the cutting damage which can reach 50 μm into the crystal volume.

## 3-2 Epitaxy

Bulk crystals of III-V compounds cannot be used for the active volume of LEDs for two reasons:

(1)  For fabrication of LEDs of all colors mixed crystals are required. These can hardly be produced in bulk form because in the compound systems of interest segregation effects occur on crystallization which yield crystals with spatially variable compositions.

(2)  The materials' quality is not sufficient for optoelectronic emitters. In particular, the nonradiative minority carrier lifetime is too short. This is most probably due to intrinsic defects like vacancies and complexes involving vacancies, for example antisite defects which act as recombination centers. A typical antisite defect in GaP consists of two Ga vacancies adjacent to a phosphorus atom occupying a gallium site. The formation of these intrinsic defects is favoured in III-V compounds

because small deviations from stoichiometry can
easily occur, and because the vapor pressure of the
group V component is very high.

Therefore the active volume of an LED has to be produced by
an epitaxial method which allows growth of mixed crystals of
any composition, and which is carried out at a temperature
much below the melting point of the compound in question,
where far fewer intrinsic defects are formed.

In silicon technology on the other hand epitaxy is used merely for fabrication of structures which cannot be realized
otherwise, for example low doped layers on top of highly doped regions. Epitaxial layers of silicon are generally of
lower quality than Si bulk crystals.
With III-V compounds for use in optoelectronics, however,
epitaxy is necessary for practically all devices to produce
material of the required composition and quality.

## 3-2-1 Liquid phase epitaxy

The historically first and still most important epitaxy method for LED fabrication is liquid phase epitaxy (LPE) which
yields brighter LEDs than the other epitaxial techniques. As
indicated by the name the material to be deposited is contained in a liquid, mostly a metal-rich solution. For deposition
of III-V compounds, the most natural choice of a solvent is
the group III constituent of the compound, for example gallium in the case of GaAs. The advantage of using the group III
constituent is that this solvent does not introduce undesired
impurities. In addition, the group III metals Al, Ga, In have
low melting points and low vapor pressures, so that they can
be handled quite easily. The basic information required to
understand LPE is contained in phase diagrams which are plots
of characteristic temperatures against composition.

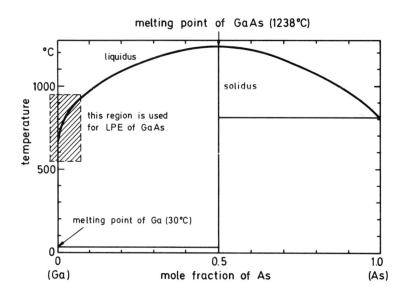

Fig. 3-4      Phase diagram of the Ga-As system

To start with LPE of GaAs, the phase diagram of Ga-As is
shown in Fig. 3-4. The liquidus line in the left part of the
figure gives the solubility of As in a Ga-As melt. If a mixture of 95 % Ga and 5 % As is heated to a temperature in excess of 880 °C (i.e. above the liquidus line) the arsenic
dissolves completely in the gallium melt. On cooling down
below the liquidus this 95/5-solution precipitates material
with the composition of the corresponding solidus which is a
vertical line at 50 % As, i.e. GaAs in this case. Due to the
precipitation of GaAs containing equal amounts of gallium and
arsenic, the solution is depleted in As and the composition
is shifted towards the Ga-rich side of the phase diagram.
This process continues on further cooling until the solution
is practically pure gallium at temperatures below 600 °C.
If this precipitation process is to be used for epitaxy, it
has to be carried out in a controlled manner so that the GaAs
is deposited in single crystalline form on a substrate. Be-

cause the epitaxial layer contains more arsenic than the solution, As has to be transported through the solution to the
solid-liquid interface. If the solution is thin (which is the
case in most practical applications of LPE) diffusion of As
in Ga is the only transport process available. Therefore the
solution has to be cooled slowly, so that As can diffuse to
the substrate before supercooling causes spontaneous nucleation within the solution. The diffusion constant D of As in
Ga at 800 °C is about $10^{-4}$ $cm^2/s$. From diffusion theory it
follows that the distance a particle can diffuse within time t
is of the order of $(Dt)^{1/2}$. If one assumes a cooling rate of
1 °C/min and a maximum supercooling of 5 °C, then the time to
reach supercooling is 5 min or 300 s, corresponding to a distance of $(10^{-4}$ $cm^2/s$ x 300 s$)^{1/2} \approx 0.17$ cm. Therefore the solution should not be thicker than roughly 2 mm in our example
to avoid spontaneous nucleation. This estimation is verified
experimentally.

It has been found by growth rate measurements that the growth
rate in liquid phase epitaxy is diffusion limited. This means
that the attachment of the atoms to the crystal lattice is
much faster than the diffusion transport, or in other words
that thermodynamical equilibrium is attained at the surface.

Phase diagrams of ternary systems like Ga-Al-As for deposition of $Ga_{1-x}Al_xAs$ are naturally much more complicated than
those of binary systems. Fig. 3-5 gives some data for the
solubility of As in Ga-Al melts with small Al contents of up
to 5 %. As can be seen from the set of curves, the solubility
decreases very much with increasing Al content. This means
that for a given layer thickness of $Ga_{1-x}Al_xAs$ much more solution must be provided and/or a larger temperature interval
has to be used than for a GaAs layer of the same thickness.
Another complication present in the Ga-Al-As system is not
evident from Fig. 3-5: the material grown from this ternary

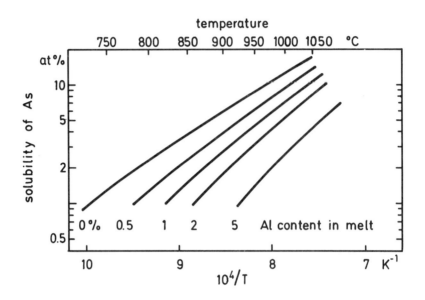

Fig. 3-5      Solubility of As in Ga-Al melts

solution exhibits a very much higher aluminum content than
the solution. The factor describing this segregation effect
is called distribution coefficient k, defined as the ratio of
concentration in the solid to concentration in the liquid. In
the case of $Ga_{1-x}Al_xAs$ the distribution coefficient of Al is
about 100 (depending on temperature or As concentration respectively). Therefore only very small Al concentrations in
the solution are required to obtain the Al mole fractions
required in LEDs which are mainly in the range of 10 to 70 %.
The high k value causes a very rapid depletion of aluminum in
LPE of $Ga_{1-x}Al_xAs$, a fact which has to be kept in mind in the
design of an LPE device. Only in very thin layers as used in
laser structures can the Al content be considered to be practically constant.

The same effects also occur in the other ternary and quaternary systems. $In_{1-x}Ga_xAs_{1-y}P_y$ for example, which is used for

emitters and detectors in the 1.3 to 1.7 μm wavelength region, has to be grown with exact lattice matching to the InP
substrate to achieve high quality layers. Therefore the composition must be adjusted exactly and kept approximately constant during epitaxy which is only possible for thin quaternary layers. Generally the difficulties in liquid phase epitaxy of mixed crystals increase with increasing differences
between the lattice constants and the enthalpies of formation
of the binary constituents, the latter difference affecting
the distribution coefficients. From the ternary compounds
mentioned in chapter 2 $In_{1-x}Al_xP$ is the worst case in this
respect: both lattice constants and enthalpies of formation
of InP and AlP differ considerably.

For the practical implementation of liquid phase epitaxy many
different configurations are used. Three important examples
of LPE methods are described in the following sections: the
vertical dipping, the horizontal slider and the melt-back
method. These methods are explained by means of processes
which are actually used in LED production.

The vertical dipping method is mainly applied for LPE-proces-
ses where only one melt is necessary. An example of this type
of process is LPE of Si-doped GaAs for infrared emitting
diodes as shown in Fig. 3-6.

The Ga melt is saturated with As (which is added in the form
of GaAs) and doped with silicon and is contained in a crucible which is placed in a vertical furnace. The process am-
bient is usually high purity hydrogen. Several GaAs substrates are fixed in a holder which can be moved up and down. The
solution is first heated to 850 °C and homogenized, then the
substrates are dipped into the melt, the temperature is slowly lowered to 730 °C and the substrates are pulled out again
(see right part of Fig. 3-6). Because the amphoteric dopant

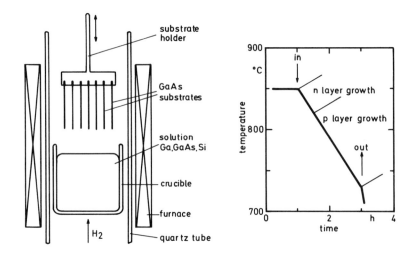

Fig. 3-6      Vertical dipping process for LPE of
              GaAs:Si, left: apparatus, right: temper-
              ature program

silicon is incorporated preferentially as a donor at temperatures above 820 °C and as an acceptor below 820 °C under
these conditions, the resulting GaAs layer grows first n type
and then p type. Thus a pn junction is grown with a single
solution. This simple and efficient process utilizes the particular properties of silicon and can therefore only be used
for growth of GaAs:Si and $Ga_{1-x}Al_xAs:Si$.

The horizontal slider method is applied for more complicated
structures with any dopants. The basic configuration is shown
in Fig. 3-7 using two-layer LPE of $Ga_{1-x}Al_xAs$ for red LEDs as
an example. The two solutions (each consisting of Ga, Al, and
GaAs for saturation, plus Zn and Te respectively as dopants)
are contained in separate wells of a slider. The GaAs substrate is placed in a recess in the base plate which is made

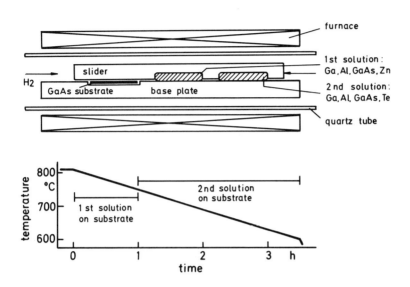

Fig. 3-7     Horizontal slider method for LPE of
             $Ga_{1-x}Al_xAs$, top: apparatus, bottom:
             temperature program

of high-purity graphite, as is the slider. The epitaxy process is performed in the following way: the furnace is heated
to a temperature so that the first solution is just saturated. Then the first solution is brought into contact with the
substrate and controlled cooling of the furnace is started.
When the Zn-doped p layer is grown to the required thickness,
the slider is moved further, separating the first solution
from the substrate and replacing it by the second solution,
the composition of which is adjusted to be saturated at the
actual temperature at this time. Further cooling results in
growth of the second layer which is n type due to the Te-dop-
ing. Finally the growth is terminated by another motion of
the slider. This method can also be extended to growth of
multilayer structures as required for example for semiconductor lasers.

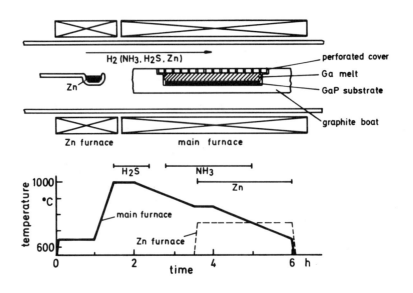

Fig. 3-8    Melt-back LPE of GaP with doping from the
            vapor phase, top: apparatus, bottom:
            temperature program

The melt-back method was developed mainly for fabrication of
green LEDs made from GaP (Fig. 3-8). Here the GaP substrate
lies in a graphite boat and is covered by the Ga melt, which
is held in place by a perforated cover. The boat is placed
inside a furnace in a gas stream of hydrogen. Doping is accomplished via the vapor phase with gaseous $NH_3$, $H_2S$ and $Zn$.
An evaporation source of elemental Zn with a separate furnace
is provided for the production of Zn vapor. At the beginning
of the process the melt is brought into contact with the substrate without further additions. After a wetting and cleaning period at 650 °C the temperature is raised to about
1000 °C. This results in dissolution of a 50 µm thick layer
from the substrate surface. When the boat is cooled down
again, the dissolved GaP grows as an epitaxial layer upon the
substrate. During this regrowth process dopants are supplied

to fabricate the desired structure: the first grown material
is n type due to the addition of $H_2S$ to the gas stream, which
leads to sulfur doping. After growth of the n layer, which is
only lightly doped towards the end of growth, the cooling is
interrupted to achieve an abrupt junction and Zn vapor is
introduced by heating the Zn furnace. Further cooling results
in a p layer on top of the n layer. The isoelectronic center
nitrogen is incorporated in both n and p layer by addition of
ammonia $NH_3$ to the carrier gas. To minimize reabsorption of
light in the devices, the nitrogen is introduced only in the
vicinity of the pn junction.

For economic mass production of LEDs liquid phase epitaxial
processes are performed on typically 20 to 50 substrates simultaneously, equivalent to several hundred thousands of
LEDs.

## 3-2-2 Vapor phase epitaxy

While liquid phase epitaxy relies on precipitation from a
saturated solution, in vapor phase epitaxy (VPE) the material
is deposited from the gas phase using suitable chemical reactions. This type of process has a larger degree of flexi-
bility than LPE because some of the restrictions imposed by
the phase diagrams do not exist here. This makes VPE especially suitable for the growth of mixed crystals. In LED
technology VPE is mainly used for the deposition of
$GaAs_{1-x}P_x$. This process is schematically shown in Fig. 3-9.

The group V components are introduced in form of the gaseous
compounds $AsH_3$ (arsine) and $PH_3$ (phosphine). The volatile
compound GaCl is formed in the source zone by reaction of
gallium metal with hydrogen chloride (at high temperatures
the monochloride GaCl is formed rather than the better known
trichloride $GaCl_3$):

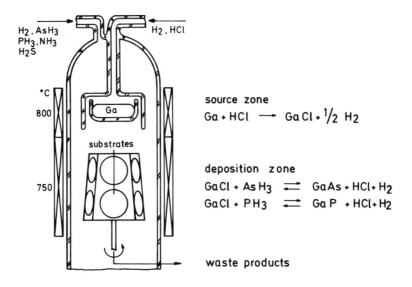

Fig. 3-9      Vapor phase epitaxy of $GaAs_{1-x}P_x$ in a
vertical barrel reactor

$$Ga + HCl \longrightarrow GaCl + 1/2\ H_2 \qquad (3-1)$$

In the deposition zone gallium chloride reacts at somewhat
lower temperature to form $GaAs_{1-x}P_x$, which is deposited on
the substrates:

$$GaCl + (1-x)AsH_3 + xPH_3 \rightleftarrows GaAs_{1-x}P_x + HCl + H_2 \quad (3-2)$$

The crystal composition x can be adjusted by variation of the
ratio of the $PH_3$ and $AsH_3$ partial pressures. Because the lattice constants of $GaAs_{1-x}P_x$ for red to yellow LEDs differ
considerably from the lattice constants of GaAs and GaP, a
layer with varying composition has to be introduced between
substrate and active layer. In this graded layer large numbers of dislocations are formed due to the lattice mismatch;
these dislocations are mostly bent sideways, however, so that

they do not propagate into the active layer. To minimize the lattice mismatch problem, material with phosphorus contents less than 50 % is deposited on GaAs (GaAs$_{0.6}$P$_{0.4}$, red) and compositions with more than 50 % phosphorus are grown on GaP (GaAs$_{0.4}$P$_{0.6}$, orange, GaAs$_{0.15}$P$_{0.85}$, yellow and GaP, green). Because the P-rich compositions are already in the indirect range, these materials are doped with nitrogen from gaseous NH$_3$, which is introduced into the reactor during growth of the active region. All layers are grown n type by doping with e.g. sulfur from H$_2$S. Experience has shown that it is hardly possible to form grown-in pn junctions during VPE because low n doping cannot be achieved after the reactor has been contaminated with zinc for p type doping.

Vapor phase epitaxy is especially well suited for large volume production due to its high capacity and short process time. It is also economically advantageous that all LED colors from red to green can be produced with one kind of process. Because the brightness of liquid phase epitaxial LEDs from Ga$_{1-x}$Al$_x$As (red) and GaP (green) is far superior to VPE devices, only low-price red and green LEDs are fabricated by VPE. For orange and yellow diodes, however, VPE is the only practicable production method today.

An inherent problem in VPE-grown GaAs$_{1-x}$P$_x$:N is inhomogeneity with respect to composition x and nitrogen concentration which results in color and brightness variations. This is due to changes in gas composition and temperature along the gas flow direction in a VPE reactor. Only by careful control of temperature, temperature gradient, gas composition and gas flow can this problem be minimized.

Further disadvantages of VPE are the dangerous properties (toxicity, explosiveness) of source gases and waste products, requiring stringent safety measures.

## 3-2-3 Other techniques

There are other techniques available for epitaxial growth of
III-V compounds which, however, are not used presently for
LED fabrication. One of these is metal-organic vapor phase
epitaxy (MOVPE), sometimes also denoted OMVPE or MOCVD. It is
a variety of vapor phase epitaxy which utilizes metal-organic
compounds like trimethyl-gallium $(CH_3)_3Ga$ as group III sources instead of the Ga-GaCl reaction used in standard VPE. The
principle of this method is shown in Fig. 3-10. Trimethyl-
gallium vapor is picked up by the hydrogen carrier gas in the
bubbler, mixed with arsine and reacts at the substrate surface to form GaAs:

$$(CH_3)_3Ga + AsH_3 \longrightarrow GaAs + 3\ CH_4 \hspace{2cm} (3-3)$$

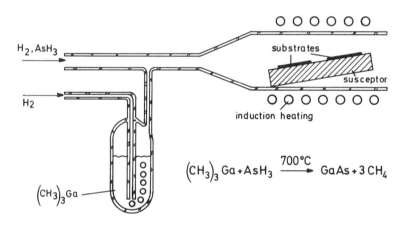

Fig. 3-10    Growth of GaAs by MOVPE, using $(CH_3)_3Ga$
and $AsH_3$

The substrate is placed on an inductively heated susceptor, so that the reactor walls remain cold, otherwise the GaAs formation reaction would also take place there. An important difference to standard VPE, which can be described by chemical equilibria, is that MOVPE uses a pyrolitic decomposition reaction which is kinetically controlled. In practice the growth rate is determined by the supply of $(CH_3)_3Ga$ only, while an excess of $AsH_3$ is present. Due to the one-way nature of the reaction, growth of mixed crystals is easily possible, even if compounds with widely differing properties are to be mixed, as for example InP and AlP. MOVPE also allows growth of Al containing compounds from the vapor phase which is difficult to do with VPE because aluminum chloride attacks quartz which is normally used for reactor construction. Although this process works at rather low temperature, yields very homogeneous layers with excellent surface quality, is suited for a very wide class of compounds and has a high production capability, it cannot be used for LED fabrication because the quality of MOVPE grown material is at the present time quite insufficient for this purpose. Instead, it is an excellent growth method for devices which do not depend critically on minority carrier lifetime, for example field effect transistors and lasers. With the onset of stimulated emission in semiconductor lasers the (low excitation) minority carrier lifetime becomes irrelevant.

Another advanced epitaxy method which is not used for LEDs is molecular beam epitaxy (MBE), where the components are evaporated separately in ultra-high vacuum and arrive simultaneously at the surface of the substrate to form an epitaxial layer (see Fig. 3-11).

The fluxes of the constituents are adjusted by the respective source temperatures. For abrupt changes in composition shutters are provided which can interrupt the molecular beams. At

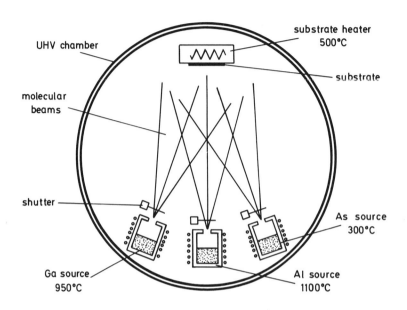

Fig. 3-11   Principle of MBE for growth of $Ga_{1-x}Al_xAs$

substrate temperatures around 500 °C typical growth rates of
0.5 μm/hour are achieved.

MBE is a very flexible method especially suited for very thin
layers because the typical growth rate is much smaller than
with other epitaxy methods. Due to the low growth temperature
very abrupt junctions can be achieved. This feature is particularly valuable for growth of highly sophisticated structures like multiple quantum well lasers and similar applications. For LED fabrication MBE offers hardly any advantage
over standard methods and is therefore not used, simply for
economic reasons.

## 3-3 Wafer processing

Wafer processing comprises all steps which are necessary to
fabricate an LED-chip from a semiconductor wafer usually covered by one or more epitaxial layers. The following subsec-
tions describe these steps; however, not all of them are ne-
cessarily needed to produce a specific type of LED. The LED
chip, displayed on the left side of Fig. 4-1 for example, may
be fabricated by the following processes: passivation, definition of the diffusion window, diffusion, metallization of
the diffused front side, definition of the front contact,
thinning and etching of the reverse side, metallization of
the reverse side, probing, i.e. functional testing on the
semiconductor wafer, dicing. This short overview shows how
specific steps have to be picked and integrated into a complete series of processing steps.

### 3-3-1 Passivation

Surface passivation serves many purposes in LED fabrication.
It is used as a diffusion barrier when defining diffusion
windows. It helps to prevent surface decomposition in diffusions and other high temperature processes. It isolates bonding pads from the semiconductor. It may be used as antireflective coating or as part of high reflective contacts.
Finally it may be used as protective coating against chemical
or mechanical attack during otherwise damaging fabrication
steps.

Materials which are used for passivation are $SiO_2$, phosphorus
doped $SiO_2$, $Si_3N_4$ and $Al_2O_3$. These are all electrically iso-
lating, stable at high temperature, chemically relatively
inert, of relatively high hardness and optically transparent
in the visible and near infrared. Characteristic data are
summarized in Table 3-2. These data are considered as typi-

Table 3-2
Characteristic data of materials used as passivation layers

|  | $SiO_2$ | $SiO_2:P_2O_5$ | $Si_3N_4$ | $Al_2O_3$ |
|---|---|---|---|---|
| Resistivity/$\Omega$m | $> 10^{13}$ | $> 10^{13}$ | $> 10^{13}$ | $>10^{13}$ |
| Dielectric strength / V/m | $5 \times 10^8$ | $10^8$ | $10^9$ | $2 \times 10^7$ |
| Dielectric constant (low freq.) | 3.5 | 4 | 6.5 | 8 |
| Refractive index | 1.455 | 1.50 | 1.9 | 1.65 |
| Thermal expansion coefficient/°C$^{-1}$ | $0.6 \times 10^{-6}$ | $1-10 \times 10^{-6}$ | $5 \times 10^{-6}$ | $7 \times 10^{-6}$ |
| Melting/softening temperature/°C | 1300 | 1000 | 1500 | 2000 |
| Chemical permeability | high | high to low | low | low |
| Hardness (Mohs) | 6 | 5 | 9.5 | 9 |

cal. In practice they show a relative large scatter, which depends on the deposition method and deposition conditions such as temperature.

Looking through the table we find that all materials are of sufficiently high electrical resistivity and dielectric strength to be suitable for our purposes. The dielectric constant of a passivation layer is unimportant in standard LED designs; however, when bonding pads are isolated from the semiconductor the capacitance of these may have to be considered in high frequency applications.

The optical refractive index of a passivating layer determines the effectiveness as an antireflective coating. Minimum

reflectivity is reached if the thickness d is adjusted to a
quarter wavelength within the material, i.e. if $d = \lambda/4n$ and
if the refractive index n is $n = (n_1 n_2)^{1/2}$, where $n_1$ is the
refractive index of the semiconductor and $n_2$ is that of the
surrounding medium. For the combination GaAs and air the indices are $n_1 = 3.6$ and $n_2 = 1$; therefore $n = 1.9$. Thus for
antireflective coatings on GaAs the most favourable choice is
$Si_3N_4$.

A thermal expansion coefficient matched to the one of the
semiconductor is favourable, because all III-V based devices
are sensitive to mechanical stress within the active volume
under forward bias, as will be described in more detail in
section 6-4. Also the diffusion coefficient of Zn and other
dopants is influenced by stress. The thermal expansion coefficients of GaAs and GaP, for example, are $5.9 \times 10^{-6}$ $K^{-1}$ and
$4.7 \times 10^{-6}$ $K^{-1}$ respectively (see Table 3-1). Therefore the
best choices for these materials are $Al_2O_3$ and $Si_3N_4$ as well
as $SiO_2:P_2O_5$ with adjusted P concentration. Matching the
thermal expansion coefficients is of highest importance in
high radiance emitters and lasers because these are very sensitive to mechanical stress.

The high temperature stability of all passivation materials
is sufficient for III-V processes like epitaxy, diffusion and
contact alloying whose temperatures are usually below
1000 °C.

Chemical permeability in the last but one row of Table 3-2
means the permeability of the passivation layer to diffusing
species in high temperature processes. For example, a diffusion of Zn through a diffusion window into a semiconductor
requires a sufficient difference between the diffusion coefficients of the passivation layer and the semiconductor. The
necessary difference depends on the depth of the diffusion

and can be relatively small when only shallow diffusions of a
few microns are to be performed. On the other hand also a
high permeability of a passivation layer may be favourably
used. Under specific diffusion conditions the semiconductor
surface can be principally unstable as explained in section
3-3-3 or can be unstable due to an unintentionally introduced
temperature gradient. In these cases a passivation of the
whole semiconductor surface by a penetrable material such as
$SiO_2$ prevents or reduces decomposition of the surface.

Finally hardness, a measure for resistance to mechanical
damage, is a parameter to be considered when a layer is used
to protect the semiconductor mechanically. An example is the
protection of the semiconductor front surface during the
mechanical thinning of the reverse side.

We turn now to methods by which III-V compounds can be passivated. Unlike on silicon a native oxide cannot be grown at
high temperatures on these compounds. Oxides can be grown
electrochemically; however, they are not of sufficiently high
thermal and chemical stability, and tend to cracks and bad
adhesiveness. Thus methods have to be used which transport
material to the semiconductor surface. Methods which are
actually employed are

> chemical vapor deposition
> plasma coating
> sputter coating and
> evaporation.

In Table 3-3 materials are displayed which can be favourably
deposited by the various methods. The main criterions to
judge a method-material combination are availability of
starting materials and deposition equipment as well as quality of the resulting passivation layer.

Table 3-3
Methods which are used to deposit passivation layers on
III-V compounds

|  | $SiO_2$ | $SiO_2:P_2O_5$ | $Si_3N_4$ | $Al_2O_3$ |
|---|---|---|---|---|
| Chemical vapor deposition | standard | standard | standard | possible |
| Plasma coating | possible | possible | standard | no |
| Sputter coating | standard | no | possible | standard |
| Evaporation | possible | no | possible | possible |

<u>Chemical vapor deposition</u> (CVD) is probably the most commonly
used method to passivate III-V compounds. The process is a
thermally activated oxidation mostly of hydrogen compounds;
therefore it is more specifically called pyrolytic deposition
or pyrolysis. Beside hydrogen compounds a large variety of
other compounds can also be used. The passivating materials
may be formed from hydrogen compounds according to the following chemical reactions:

$SiO_2$ :

$$SiH_4 + 2O_2 \xrightarrow{> 350 \ °C} SiO_2 + 2H_2O \qquad (3-4)$$

$SiO_2:P_2O_5$ :

$$SiH_4 + 2O_2 \xrightarrow{> 350 \ °C} SiO_2 + 2H_2O \qquad (3-5)$$

$$2PH_3 + 4O_2 \xrightarrow{> 300 \ °C} P_2O_5 + 3H_2O \qquad (3-6)$$

$Si_3N_4$ :

$$3SiH_4 + 4NH_3 \xrightarrow{> 400 \ °C} Si_3N_4 + 12 \ H_2 \qquad (3-7)$$

A similar reaction for the formation of $Al_2O_3$ does not exist; however, it can be formed from aluminum isopropoxide, $Al(OC_3H_7)_3$. In spite of its attractive properties $Al_2O_3$ is less frequently used for passivation purposes than the other passivating oxides.

The pyrolitic decomposition of the gaseous compounds silane, $SiH_4$, and phosphine, $PH_3$, becomes practicable at the temperatures indicated in the reaction formulas. Typical process temperatures are, however, higher. Amorphous glassy silicon dioxide is formed at 400 °C to 700 °C; within the same temperature range phosphorus doped silicon dioxide $SiO_2:P_2O_5$ or phosphosilicate glass is obtained. Pure phosphorus pentoxide $P_2O_5$ is strongly hygroscopic; therefore the phosphosilicate glass is usually covered by a layer of undoped silicon dioxide. By chemical vapor deposition this can be very easily accomplished. The typical process temperature for silicon nitride formation is between 550 °C and 900 °C. The lower process temperatures within the ranges indicated are always to be preferred for III-V compounds.

A deposition reactor is schematically displayed in Fig. 3-12. The reaction gases are diluted in an inert gas like nitrogen, hydrogen, argon or helium. To prevent premature reactions oxygen is usually fed directly into the reaction furnace whereas the other compounds are thoroughly mixed before entering. The thermally activated deposition yields highly uniform coating films. The deposition rate depends on deposition temperature, concentration of reactants, gas flow rates and geometry of the reaction zone. Typical deposition rates are 5 to 20 nm/min.

The films produced by chemical vapor deposition can be of very high quality. The adhesion is excellent if clean wafers are used. The thickness uniformity is usually good. Uniformi-

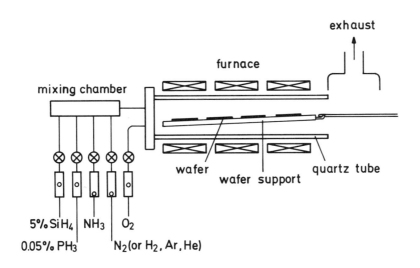

Fig. 3-12   Schematic representation of a horizontal
            CVD reactor for the deposition of $SiO_2$,
            $SiO_2 : P_2O_5$, and $Si_3N_4$.

ty may be qualitatively evaluated by looking at the inter-
ference color of the deposited layer. Thickness variations of
about 5 % are easily recognized. The control of the absolute
thickness of the films can be sometimes difficult. Variations
of $\pm$ 15 % are typical. Stoichiometry of the film is usually
excellent judged by the measured refractive index. Refractive
index as well as thickness can be measured at the same time
by the ellipsometric method.

The density of the films is high due to the relatively high
deposition temperature. On the other hand pin hole and crack
densities are low. Therefore the passivating properties of
the layers are attractive for diffusions and alloying. Finally the high deposition temperature of CVD films, high for

III-V compounds, can be a disadvantage when temperature sensitive wafers are to be coated, for example wafers with alloyed contacts.

The equipment used for chemical vapor deposition is readily
available from the suppliers of the semiconductor industry.
However, safety precautions have to be elaborate because most
of the reactants used are extremely toxic, explosive, or cor-
rosive.

The plasma deposition method is in some respects very similar
to the CVD method. Like CVD it is a purely chemical process
using the same or similar gaseous compounds as CVD. Unlike
CVD, however, the energy needed to promote the reaction of
the gases is provided by energetic electrons of the plasma
instead of by heat. Therefore, the temperature of the substrates can be as low as room temperature and this is the
main advantage of the reactive plasma deposition method. Reactions which are exploited can be the same as indicated in
Eq. 3-4 to 3-7. As in CVD the gases are diluted in an inert
carrier gas, for example nitrogen or argon.

In Fig. 3-13 the voltage current characteristic of a gas discharge is displayed at a pressure in the range $10^{-3}$ mbar to
1 mbar. There are four different types of discharges:
Townsend discharge, normal and abnormal glow discharge and
arc discharge. The transitions between the discharge regimes
depend on the pressure through the free path lengths of ions
and electrons, on the geometry of the discharge vessel and
the nature of the gases through their ionization coefficients. Independently of the specific configuration the types
of discharges can, however, be characterized by the dominating physical processes. In the regime of the Townsend discharge the current is sufficiently low for not yet being
distorted by space charges. The voltage increases only weakly

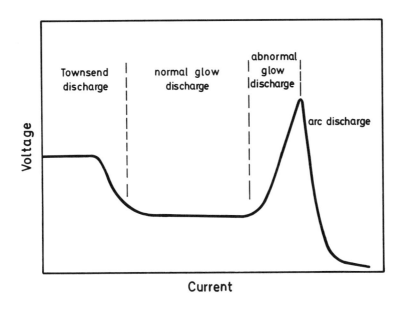

Fig. 3-13    Voltage-current characteristic of a d.c.
             gas discharge

with the current, because the moving charge carriers, preferentially electrons, are multiplied by impact ionization.
Beyond a certain current level the multiplication tends to
produce more charge carriers than disappear at the electrodes. A steady state can be maintained only by the build up of
space charges, important in all other types of discharges.
The normal glow discharge is characterized by a constant
voltage at which the current can be increased over several
orders of magnitude. This is accomplished by the adjustment
of the current carrying area of the cathode. Only when the
whole cathode is covered by the discharge does the cathode
current density increase and the voltage increase into the
abnormal discharge regime.

The dependency of the electric potential from the position x
between cathode and anode of a glow discharge is shown in

Fig. 3-14. The field strength is highest in the region of the
cathode fall and very low in the region of the positive column. Changing the distance between cathode and anode does not
change the magnitude and width of the cathode fall; also the
overall voltage is hardly changed, only the length of the
positive column adjusts.

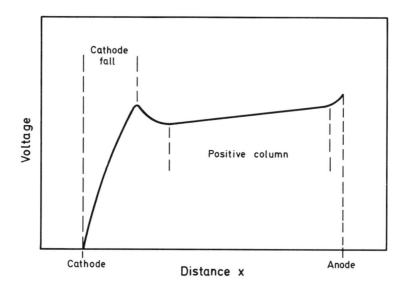

Fig. 3-14   Voltage versus position characteristic of
a glow discharge

The glow discharge is exploited for plasma deposition. The
cathode fall is the region where electrons and ions pick up
enough energy to induce chemical reactions according to Eq.
3-4 to 3-7, for example. If the positive column has not
enough space to develop, the abnormal glow regime is entered.
It is characterized by higher current density and higher
field strength in a larger volume. These are the operating
conditions of sputter deposition. At even higher current densities an arc-discharge develops due to the production of
additional electrons thermally emitted by the cathode.

A plasma deposition reactor is shown in Fig. 3-15.
Practical systems always use radio frequency power in the
range of several MHz because otherwise the electrodes would
be passivated quickly. The gaseous reactants are introduced
through a gas cabinet into the reaction chamber, where the
plasma is maintained at a pressure in the range 0.1 to
10 mbar. The plasma consists in the high field regions mostly
of ions and highly reactive radicals. The reaction product
deposits on the wafers and the surrounding electrode. The
temperature of the wafers can be controlled by a resistance
heater. Similarly as in chemical vapor deposition the deposition rate depends on the concentration of the reactants, the
gas flow rate, the geometry of the reaction chamber and the
radio frequency power. Typical deposition rates lie between 2
and 10 nm/min.

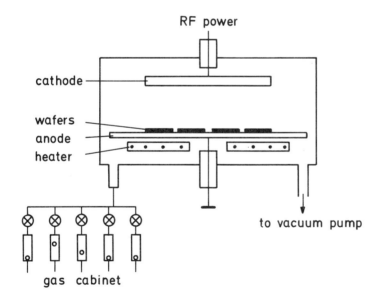

Fig. 3-15   Schematic representation of a plasma depo-
             sition reactor

There are two main advantages of the plasma deposition method. The first is the possibility of choosing deposition temperatures as low as room temperature, as mentioned above, and
the second is the availability of an etching process, as
described below.

Low temperature deposition is attractive because one does not
have to worry about temperature-induced deterioration of
semiprocessed wafers. However, usually the density of these
passivations is lower and the pin hole density higher than
that of films deposited at higher temperature. Therefore the
usefulness of very low temperature deposition has to be
judged carefully.

The etching process may be used either to clean surfaces before deposition of a passivation or to etch selectively
through a passivation film partly covered by a structured
photoresist. In both cases, the gases used depend as in wet
chemical etching on the material to be etched. Versatile
gases for etching purposes are halocarbon types, like $CF_4$,
$CFCl_3$ and others.

In spite of the advantages described, the plasma method is
not very commonly used, possibly because both advantageous
features are also offered by sputter deposition. This method
leads to better quality films at low temperature and allows
etching of arbitrary materials using just one gaseous compound which moreover can be applied for deposition.

Sputter coating is, after CVD, probably the most widely used
deposition method. As in plasma deposition a plasma is essential; however, the gas discharge is maintained in the abnormal discharge regime and therefore a much higher field
strength in front of the cathode is reached than in the normal cathode fall. The energy of the positive ions impinging

on the cathode is made large enough to cause ejection of
atoms or molecules out of the cathode, which is then called
the target. These atoms or molecules deposit on the semiconductor substrates mounted in close neighbourhood of the target. In this process the plasma ions act merely as physical
projectiles transferring impulse and kinetic energy to the
target atoms. Chemical reactions in the plasma are prevented
by using an inert gas, usually argon, without additions.

The yield $\eta_s$ of the sputter process is defined as the number
of target atoms $n_t$ which are ejected per $n_i$ impinging plasma
ions:

$$\eta_s = n_t/n_i \qquad\qquad (3-8)$$

For practical purposes the deposition rate is more important
than the sputter yield. One finds that the deposition rate is
proportional to the electrical power fed into the plasma for
most target materials. Therefore specific deposition rates
are measured in nm/kW min. These rates depend also on target
size which has to be stated but disc shaped targets with
diameters of 200 mm are fairly standard. Examples of typical
deposition rates are given in Table 3-4.

In Fig. 3-16 the schematics of a sputter system are shown.
The wafers to be coated lie on a table which electrically is
at or close to anode potential and which usually rotates during deposition. One or more targets of different materials
form the cathode. The distance target to wafer is typically
50 mm. Before each sputter process the system should be thoroughly evacuated into the $10^{-7}$ mbar range. Traces of water,
nitrogen and oxygen adversely effect the adhesiveness and
roughness of deposited films and the resistivity of conducting films is changed. After evacuation argon gas is fed into
the system to a pressure between $10^{-3}$ and $10^{-1}$ mbar.

Before deposition is started the target should be sputtered
for a short time with the wafers shielded by a shutter. In
this way deposition of pure material is ensured.

There are special features which are frequently available on
a single system. Direct current sputtering is the oldest and
cheapest method. However, it can only be used with conductive
targets, therefore passivation is not possible in this way.

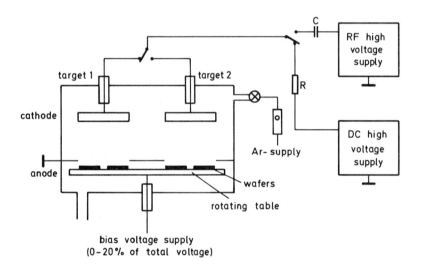

Fig. 3-16    Schematic representation of a sputter de-
             position reactor

Radio frequency (RF) sputtering, usually at 13.56 MHz, is the
solution for non-conductive targets. It can, however, also be
used for metals. The target becomes the cathode under the
influence of the RF field due to two facts. First the frequency of the RF field is in a range where the low mobility
species of the charged plasma particles, the ions, cannot
follow the field, whereas the electrons with their high mobility follow the field. Therefore a positive space charge is

built up consisting of ions. Second, as indicated in Fig.
3-17, there is a large difference between the areas of the
target and the opposing electrode. Thus the electric field is
not homogenous but smallest close to the larger area electrode. In this region the positive space charge is located.

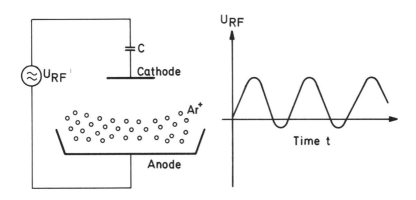

Fig. 3-17   Principle of RF sputtering. On the right
            hand side the voltage between anode and
            cathode is shown at a frequency where the
            $Ar^+$ ions cannot follow the electric field.

The capacitance between cathode and RF generator prevents the
compensation of the charge of the cathode and causes the target to become the cathode whereas no permanent charge can be
built up at the grounded anode. The time dependence of the
anode to cathode voltage is also shown in Fig. 3-17. During
most of a period the cathode is negative, causing positive
ions to sputter away the target material. At the same time
electrons bombard the wafers and may cause considerable heating of the growing film. During the rest of the period the

polarity is reversed: electrons compensate the collected positive charge on the target and positive ions impinge on the
wafers. However, without additionally biasing the wafers, the
energy of the ions is not sufficient to sputter away the already deposited layer.

Magnetron targets can be favourably used to avoid the heating
effect by electrons bombarding the wafers. The magnetron
principle is shown in Fig. 3-18. A permanent magnet behind
the target produces a magnetic field which is in large part
perpendicular to the electric field. The north pole of the
magnet completely surrounds the south pole, or vice versa.
The Lorenz force, perpendicular to the electric field and the
magnetic field, causes the electrons to spin on cycloidal
tracks around the gaps between north and south pole. Therefore most of the electrons are prevented from bombarding the
anode. Moreover, due to their much increased path length in
the plasma, the production of ions per electron is higher.
For a given input power, the sputter rate is therefore up to
a factor of five higher, or the input power may be reduced,
thus lowering the sputter rate but further decreasing the
heating effect on the wafers. Additionally, the gas pressure
can be lower due to the higher ionization and this improves
purity of deposited layers. All these advantages made magnetron sputtering the most frequently used sputter method.

Bias sputtering means that part of the DC or RF power, usually below 10 to 20 %, is applied to the support of the wafers.
In this way some argon ions are attracted to the growing film
which is partly sputtered away. Because the sputtered atoms
fly into directions having a component parallel to the surface, step coverage is much improved by bias sputtering. Not
yet always understood are the frequently observed beneficial
effects on pin hole density and reduction of stress within
films.

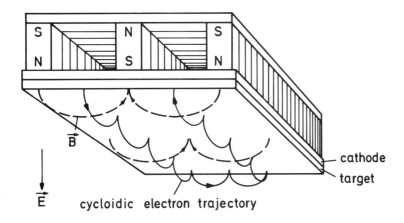

Fig. 3-18   Magnetron target with magnetically induced
cycloidic electron tracks

Sputter etching is a feature incorporated in most systems.
For this purpose the roles of target and wafer support are
interchanged causing ions to sputter the wafers. It is mainly
used to clean the wafers before depositing the film to promote adhesion. During sputter etching a shutter is positioned
in front of the target to protect it against contamination by
impurities. Sputter etching can also be used for pattern de-
lineation. However, it has to be kept in mind that all materials are sputtered, including the masking photo resist.
Therefore the differences in sputter rates and thicknesses
between the masking and the material to be sputtered have to
be considered carefully. Beside photo resists, metals with
low sputter yields like Ti or Al are utilized as masks.

In Table 3-4 a few typical specific deposition rates are summarized for RF diode and RF magnetron deposition. The rates

___

are average rates for rotating wafers which are only about 1/10 of the time under circular targets of 200 mm diameter.

Table 3-4
Typical specific deposition rates of sputter processes with rotating wafers. Targets are discs with 200 mm diameter.

| Material | Al | $Al_2O_3$ | Si | $SiO_2$ | $Si_3N_4$ | Au | Ag | |
|---|---|---|---|---|---|---|---|---|
| RF diode | 4.0 | 1-2 | 2-3 | 2-3 | 1-2 | 10.0 | 14 | nm/kW min |
| RF magnetron | 12.0 | 5-8 | 8-12 | 8-12 | 5-8 | 24.0 | 29.0 | nm/kW min |

The quality of sputter deposited $Al_2O_3$ is excellent. This is true with regard to adhesion, thickness uniformity and thickness control, stoichiometry, density and pin hole density. However, stoichiometric deposition of $SiO_2$ and $Si_3N_4$ is not straightforward. Usually a deficiency of oxygen or nitrogen is observed. This can only be prevented by adding oxygen or nitrogen to the sputter gas and turning in this way to the so called reactive sputter process. With this process not only $SiO_2$ and $Si_3N_4$ targets can be used but one may even utilize pure silicon.

The last deposition method is evaporation. The source material is heated in a vacuum below about $10^{-6}$ mbar and vaporized. Heating is usually performed by an electron beam of 5 to 10 kW power. This is a standard method for the deposition of metals and metal alloys. Deposition of dielectrics is possible but poses a series of problems. First, the melting of the source material is, due to its low heat conductivity, much more difficult than that of a metal. Second, all three materials $SiO_2$, $Si_3N_4$, and $Al_2O_3$ will become non-stoichiometric unless oxygen or nitrogen are added during vaporization. The adhesion, density, and pin hole density is best for $Al_2O_3$

but often not sufficient for $SiO_2$ and $Si_3N_4$. Nevertheless the possibility of room temperature deposition makes this method attractive for processing steps where coating quality is not of supreme importance as for example in coatings which protect temporarily against handling damage.

### 3-3-2 Pattern definition by photolithography

The principle of photolithography pattern definition is explained in Fig. 3-19. Let us suppose that a contact metallization has to be defined. At first the metallized wafer will be coated by a photo resist of uniform thickness of about 0.5 to 1.5 µm. Coating is usually done by the spin-on technique. The liquid resist is dropped on the wafer rotating at 2000 to 8000 rpm. The resist spreads evenly over the wafer and dries quickly. Final thickness depends on rotation frequency and viscosity of the resist. After an additional drying at about 90 °C the wafer is ready for exposure. For this purpose ultra-violet light of the wavelength region 300 to 400 nm shines trough a mask on the photoresist as shown in Fig. 3-19. According to the type of the resist, positive or negative, the exposed resist becomes soluble or non-soluble to the developer. After development of the positive resist the non-transparent pattern of the mask is translated into a resist structure and vice versa for the negative resist as indicated in Fig. 3-19. The resist is now baked for about 20 min at typically 120 °C. Now the metal film in our example can be etched in the areas not coated by resist. Finally the resist on the metal pattern is removed and the wafer is ready for the following process steps.

Mask fabrication is an important step in photolithography. Earlier the enlarged non-repetitive structure of the mask was cut from red plastic foil, photographically reduced, and identical parts put together by step and repeat cameras to

form the repetitive mask pattern. Today the structure of the
mask is visualized by computer aided design (CAD) on a computer screen. In this stage dimensions are easily changed,
the structure compared with earlier designs, or details added
or deleted. The final design is stored, usually on magnetic
tape, in a format which is directly readable by an electron
beam lithography system. This system produces a mask much in
the same way as shown in Fig. 3-19 with the exception that
the UV-light and the mask in the figure are replaced by an
electron beam under the control of the digitized data produced by CAD. The mask production by the electron beam system
is relatively slow, therefore only one mask, the master mask,
is fabricated in this way, whereas working masks are copied
from it according to the conventional method.

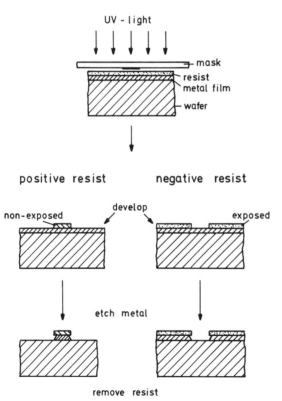

Optical exposure systems needed for the first step in Fig. 3-19 work according to one of three principles. These principles are contact, projection, or proximity exposure. In a contact system the mask is pressed against the wafer as shown in Fig. 3-19. This is the simplest method and moreover it has the advantage that bowed wafers, quite common among epitaxial III-V material, are pressed flat. A disadvantage is

Fig. 3-19  Pattern definition by photolithography

the possibility that resist may stick to the mask and may be
transferred to the next wafer, thus producing two defective
wafers. The two other methods avoid this disadvantage at the
expense of higher cost. In a projection system the image of
the mask is optically projected on the wafer, whereas in a
proximity system the mask is illuminated by a highly parallel
beam so that the mask can be moved away from the surface by
about 20 μm and still produce a well defined shadow.

## 3-3-3 Diffusion

Diffusion is a purely statistical phenomenon, meaning that an
unlikely distribution of impurities transforms into a more
likely distribution. Diffusion stops when the concentration
of impurities within a volume is uniform. Concentration gra-
dients are the driving forces behind diffusion. Therefore it
is plausible to suppose that the flux density I of diffusing
impurities, measured in atoms per $m^2$ per second, is proportional to the gradient of the concentration N:

$$I = - D \frac{\partial N}{\partial x} \qquad (3-9)$$

This is Fick's first law of diffusion. For any combination of
I and $\partial N/\partial x$ it defines the diffusion constant D. As a physical law it states that D is a constant. The minus sign is
introduced because diffusing atoms move from high to low concentration. Generalized to three dimensions Eq. 3-9 takes the
form

$$\vec{I} = - D \text{ grad } N \qquad (3-10)$$

The concept of mass continuity may be applied to Eq. 3-9, as
visualized in Fig. 3-20. The change per second of the number
of atoms within the volume AΔx is equal to the difference of
atoms which enter the volume at x and leave it at x + Δx:

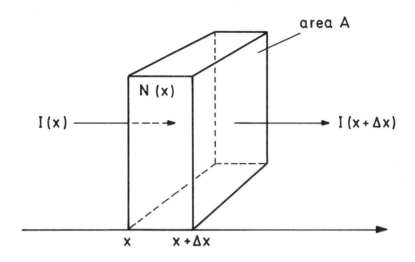

Fig. 3-20   To the derivation of Fick's second law

$$\frac{\partial N(x,\ t)}{\partial t}\ A\Delta x = AI(x) - AI\ (x+\Delta x) \qquad (3-11)$$

With (3-9), the right side of this equation is transformed to

$$I(x) - I(x+\Delta x) = - D \left\{ \frac{N\ (x)}{x} - \frac{N(x+x\Delta)}{x} \right\}$$

and with $N(x+\Delta x) = N(x) + \frac{\partial N(x)}{\partial x}\ \Delta x$   one obtains:

$$I(x) - I(x+\Delta x) = - D \left\{ - \frac{\partial^2 N(x)}{\partial x^2}\ \Delta x \right\}$$

Inserting in Eq. 3-11, one obtains Fick's second law:

$$\frac{\partial N\ (x,\ t)}{\partial t} = D \frac{\partial^2 N\ (x,\ t)}{\partial x^2} \qquad (3-12)$$

Generalized to three dimensions it reads:

$$\frac{\partial N}{\partial t} = D \text{ div grad } N \qquad (3-13)$$

If the pn junction area is large compared to the diffusion depth, the one dimensional Eq. 3-12 is appropriate. This is a linear differential equation. Solutions can be selected by defining boundary conditions. The most frequently encountered boundary condition is in-diffusion from a constant surface concentration $N_0$. The diffusing impurities enter the body, thought to be semiinfinite, through the plane x = 0 at times t > 0.

Therefore:

$$N(x,0) = 0 \text{ for } x > 0 \text{ and } N(0,t) = N_0 \text{ for } t \geq 0 \qquad (3-14)$$

The solution of (3.12) with (3.14) is:

$$N(x,t) = N_0 \left\{ 1 - \frac{2}{\sqrt{\pi}} \int_0^{\frac{x}{2\sqrt{Dt}}} e^{-z^2} dz \right\} = N_0 \text{ erfc } (\frac{x}{2\sqrt{Dt}}) \qquad (3-15)$$

as may be verified by differentation. The function erfc is the "complementary error function", known from error theory. It is displayed in Fig. 3-21 on a linear scale and, more familiarly, on a logarithmic scale as well. If a non-diffusing background impurity with concentration $N_b$ and opposite conductivity is present, the net concentration is:

$$N_{net} = N_0 \text{ erfc } (\frac{x}{2\sqrt{Dt}}) - N_b \qquad (3-16)$$

The pn junction depth $x_j$ is defined by $N(x_j) = N_b$ or $N_{net} = 0$, therefore:

$$x_j = 2\sqrt{Dt}\ \mathrm{erfc}^{-1}\ (N_b/N_0) \qquad\qquad (3\text{-}17)$$

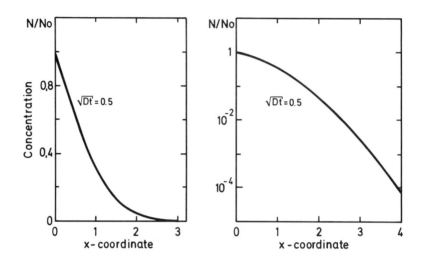

Fig. 3-21   Theoretical diffusion profile on a linear
(left) and a logarithmic scale (right).
Note, that $\sqrt{Dt}$ has the dimension of a
length, and is given in the same units as
x.

This result is extremely useful to calculate diffusion
depths. It states that the junction depth is proportional to
the square root of the diffusion time.

Other solutions of Eq. 3-12 are obtained by different boundary conditions. Examples are: out-diffusion with zero surface
concentration, or diffusion from a limited source, e.g. an
implantation of impurities.

The diffusion constand D is strongly temperature dependent.
Experimentally one finds an exponentional dependence:

$$D = D_0 \ e^{-E_a/kT} \qquad\qquad (3-18)$$

Here $E_a$ is an activation energy of the diffusion process, $D_0$
is the diffusion constant which is approached when the temperature T approaches an infinite value, and k is the Boltzmann constant.

In LED-technology exclusively p type diffusions are used.
Zinc is the most popular choice, but cadmium, also 2-valent,
is possible too. The reason is that Zn diffuses very rapidly
due to a speciality of the diffusion process: interstitial
diffusion. All n type impurities diffuse in the regular substitutional manner and are therefore slow. This is a decisive
disadvantage in III-V technology, because all these materials
tend to decompose and change stoichiometry during high temperature treatments at low pressure for extended times.

It is amazing that despite the many complications of the in-
terstital-substitutional diffusion mechanism, Zn has become
the nearly exclusively used diffusion source. The interstitial-substitutional model assumes that an impurity, e.g. zinc
in a III-V compound, exists either in the form of an interstitial atom or as a substitutional atom on the site of the
3-valent constituent. The substitutional zinc acts as the
useful acceptor, its diffusion is negligibly slow compared to
that of the interstitial zinc, but the concentration of the
interstitial species is very small compared to that of the
substitutional species. Diffusing zinc moves therefore quickly through the crystal lattice on interstitial sites until it
finds a 3-valent vacancy and occupies it as an acceptor. In
this way a vacancy has disappeared and a Zn atom has dropped
out of the rapid diffusion process. The abundance of tri-
valent vacancies thus influences the equilibrium between the
rapid and slowly diffusing zinc. Therefore the diffusion constant D will be not constant, and actually will increase with

concentration unless the crystal is able to produce vacancies
as fast as they are consumed by substitutional zinc atoms.
When not enough vacancies can be produced, the assumption of
Fick's laws of a concentration independent diffusion constant
is not valid. One consequence is that the diffusion profile
does not follow an erfc-curve. This deviation has been shown
experimentally by many authors. Fig. 3-22 presents diffusion
profiles of zinc of GaP, adapted from results of TUCK and JAY
(1977), which show pronounced deviations from an error function profile. The concave section is typical for Zn diffusion
profiles in all III-V compounds. It is believed that it is
due to the depth dependence of the dominant vacancy production mechanism. Close to the surface consumed vacancies are
subsequently replaced by indiffusion from the surface, whereas in the bulk a production mechanism of a much lower rate such as dislocation climb has to be assumed.

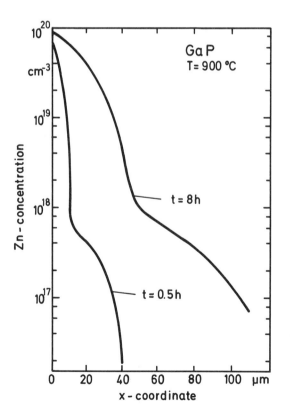

Fortunately enough the time dependence of the diffusion depth according to Eq. 3-17 is not changed by the interstitial-substitutional diffusion mechanism. Junction depth is a linear function of the square root of diffusion time, as shown in Fig. 3-23, according to data of SHIH

Fig. 3-22   Experimental Zn diffusion profiles in GaP

(1976) for GaAs. The strong temperature dependence of the
diffusion constant is reflected by the temperature dependence
of diffusion depth. Another interesting feature is the influence of the diffusion source on junction depth. $ZnAs_2$ produces a higher $As_4$ vapor pressure than $Zn_3As_2$; therefore the
concentration of As vacancies is depressed in the case of
$ZnAs_2$. Because the total number of vacancies, As plus Ga, is
constant at a given temperature for thermodynamical reasons,
an increased $As_4$ vapor pressure induces a higher Ga vacancy
concentration. According to the interstitial-substitutional
diffusion model this means a higher fraction of substitutional zinc on Ga sites and therefore slower diffusion as displayed in Fig. 3-23. An-
other consequence of the
higher $As_4$ vapor pressure
is a higher surface con-
centration of substitu-
tional zinc and therefore
an increased p type dop-
ing level.

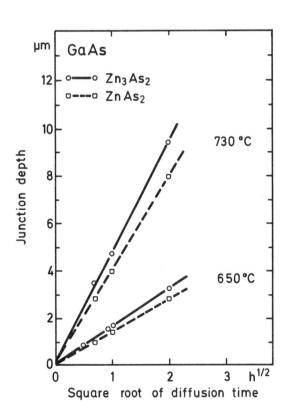

The diffusion technique
used for LED production
is quite different from
that used in silicon
technology. Whereas dif-
fusions in silicon are
performed in open tubes,
this is not possible with
III-V compounds due to
the high vapor pressure
of the V component. Al-

Fig. 3-23    Time dependence of diffused junction depths in
             GaAs

though many attempts have been made to adapt open tube con-
cepts to III-V technology, the standard method, which has
been highly perfected, is closed ampoule diffusion. In this
technique 50 or more semiconductor wafers, containing about
500 000 LED chips, are placed inside an evacuated quartz ampoule. The diffusion source is Zn, or a Zn compound producing
a vapor pressure of the 5-valent component. Reproducibility
is improved by using slightly more of the diffusion source
than vaporizes at the diffusion temperature. The selection of
the diffusion source depends on several aspects. Surface decomposition of the wafers caused by the volatile V compound
or by the formation of liquid phases within the system semi-
conductor-diffusion source poses a problem. CASEY and PANISH
(1968) pointed out that the latter problem can be overcome by
consulting the ternary phase diagram Ga-As-Zn in the case of
a Zn diffusion into GaAs. A diffusion temperature and a composition of the source are selected that do not form a liquid
phase. However, another very flexible solution is the passivation of the semiconductor surface by a material that is
transparent to diffusing zinc atoms. This technique protects
the semiconductor surface at the same time from damage by a
liquid phase and by the evaporation of the 5-valent component. An ideal passivation for this purpose is an $SiO_2$ film
with a thickness of 50 to 100 nm. This film is easily penetrated by zinc and gallium, but not by the 5-valent elements
arsenic and phosphorus.

Fig. 3-24 shows a diffusion ampoule containing semiconductor
wafers and the diffusion source. The ampoule is evacuated to
a vacuum below $10^{-6}$ mbar and sealed. Typical diffusion temperatures are 700 °C to 750 °C, and diffusion time range between one and two hours. At the end of the diffusion time the
ampoule is rapidly cooled down to room temperature. The wafers are taken out of the ampoule and are now ready for further processing.

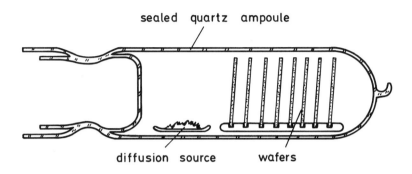

Fig. 3-24   Zn diffusion ampoule

### 3-3-4 Metallization

Metallization in LED technology has to provide ohmic contacts
to the light emitting pn junction, to form bonding pads, to
connect bonding pads with ohmic contacts and last but not
least to form contacts with high optical reflectivity. Means
to reach the latter aims are described in chapter 4, "Types
of LEDs"; here we are concerned only with the formation of
ohmic contacts. An ohmic contact should have a linear current-voltage characteristic over its whole current range.
Such an ideal behavior rarely exists either in theory or in
semiconductor technology. From a practical point of view, a
satisfactory ohmic contact is one that does not significantly
interfere with LED performance. That means its voltage drop
over the whole current range should be smaller than the voltage drop at the pn junction and within the bulk semiconductor.

Let us first consider ohmic contacts from a theoretical point
of view. Quasi free electrons in a metal occupy nearly all
states up to the Fermi energy $E_F$ whereas the states above $E_F$
are essentially empty at low temperature. The Fermi energy is
lower than the lowest electron energy in the vacuum; therefore an energy, called the work function, is necessary to
remove an electron from the metal. This situation is visualized in part (a) of Fig. 3-25. If a semiconductor is brought
into contact with the metal, as shown in part (b) of Fig.
3-25, then in a similar fashion, the energies of the electrons in the semiconductor differ from those in the metal.
Usually the conduction band energy of the semiconductor is
higher than the Fermi energy in the metal. Consequently electrons flow from the conduction band of the semiconductor into
the electron reservoir of the metal, the Fermi sea. These
electrons cannot fill up the Fermi sea, but the missing electric charge decreases the energy of the conduction band of
the semiconductor until the Fermi energy of the semiconductor
adjusts to that of the metal. As shown in part (c) of Fig.
3-25, a barrier is formed at the interface between metal and
semiconductor. The space charge connected with the barrier is
formed by ionized impurities which are no longer neutralized
by mobile electrons.

Assuming a uniform distribution of ionizable impurities,
Poisson's equation yields a parabolic energy barrier:

$$\emptyset(x) = e^2 N x^2 / 2 \, \varepsilon_s \, \varepsilon_0 \qquad\qquad (3-19)$$

Here e is the electronic charge, N the ionized impurity con-
centration, $\varepsilon_s$ the static dielectric constant and $\varepsilon_0$ the permittivity of free space. As shown in Fig. 3-25, (c), the band
bending $E_b$ is equal to the barrier height $\emptyset_b$ diminished by
the energy difference $\emptyset_F$ between conduction band and Fermi
level in the semiconductor. A voltage V, applied to the in-

terface, can modulate the band bending:

$$E_b = \emptyset_b - \emptyset_F - eV \qquad (3-20)$$

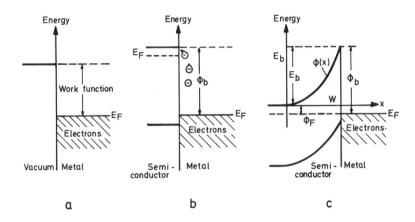

Fig. 3-25   Energy diagrams of a metal in contact with
            the vacuum (a), in the moment of touching
            a semiconductor (b), and when equilibrium
            has been reached and a barrier is formed
            (c).

According to Eq. (3-19) the band bending is also connected to
the barrier width w:

$$E_b = e^2 \ N \ w^2/2 \ \varepsilon_s \ \varepsilon_0 \qquad (3-21)$$

The barrier height $\emptyset_b$ depends on the specific combination
metal-semiconductor. However, the influence of the metal is
very weak in predominatetly covalent semiconductors like the
III-V compounds. Thus the barrier height depends in III-V
compounds essentially only on the semiconductor. Experimen-

tally one finds that the barrier is about 2/3 of the band gap
energy in n type III-V compounds and about 1/3 of the band
gap in p type material. The band bending $E_b$ can be decreased
or increased by an external voltage whereas the barrier $\emptyset_b$
cannot be changed. The consequence is that electrons can flow
from the semiconductor into the metal under a bias which decreases the band bending but not from the metal back into the
semiconductor over the barrier $\emptyset_b$. A diode characteristic
results from this, exactly of the same form as the characteristic for a pn junction:

$$I = I_0 \ (\exp \ (eV/kT)-1) \qquad\qquad (3-22)$$

Eq. 3-22 characterizes not an ohmic contact but a rectifying
Schottky contact. The emission of electrons from the semiconductor into the metal under forward bias is called thermionic
emission, because only electrons of the highest thermal energy are able to pass over the barrier. Fortunately there is
the quantum-mechanical tunnel effect, which allows electrons
to penetrate sufficiently thin energy barriers. According to
Eq. 3-21 the barrier width is proportional to $1/\sqrt{N}$ because
the band bending $E_b$ is essentially independent of the doping
level N. That means that by increasing the doping level, the
barrier width can be decreased. At first electrons can tunnel
through only close to the top of the barrier and therefore
thermal energy is necessary to reach that point. This mode of
current transport is temperature dependent and is referred to
as thermally assisted tunneling or thermionic field emission.
As the doping level is further increased, the barrier becomes
so thin that an appreciable number of electrons can tunnel
through even at the base of the barrier. This temperature
independent mode of current transport is called field emission tunneling. The current voltage characteristic is symmetrical because the barrier can be penetrated from both sides
in the same way. Field emission is the preferred mode of cur-

rent transport in ohmic contacts. The transistion between the
characteristic of a rectifying Schottky contact and an ohmic
tunnel contact is shown in Fig. 3-26.

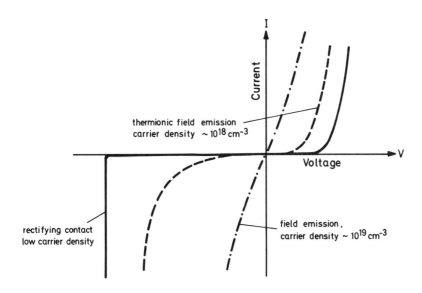

Fig. 3-26   Voltage-current characteristic of a metal-
semiconductor contact shifting from recti-
fiying to ohmic behavior

The theoretical aspects of semiconductor contacts show the
direction in which low resistivity ohmic contacts have to be
sought. There has to be a highly doped region of the semiconductor directly adjacent to the contacting metal. Such a
highly doped layer can be formed by crystalline regrowth during alloying, or by diffusion of an appropriate dopant contained in the metallization, or by a combination of both. The
most frequently used contact for n type GaAs and GaP is an
alloyed Au:Ge contact. As a typical example of an alloyed
contact, the fabrication of the Au:Ge contact is described
here in more detail.

The clean semiconductor wafers are at first metallized, preferentially by sputter coating or evaporation, processes that
have been described for dielectrics in section 3-3-1. After
deposition the film has to be alloyed. This is done in an
open tube furnace usually in a forming gas (85 % nitrogen,
15 % hydrogen) atmosphere. Upon heating the film melts close
to the eutectic temperature of 336 °C and some semiconductor
material is dissolved in the liquid. Because the group V component easily evaporates the heating cycle should be as short
as possible; usually it lasts only few minutes. Regrowth of
the semiconductor starts upon cooling of the liquid alloy. It
is believed that gold getters gallium and therefore produces
gallium vacancies in the semiconductor. These can be filled
by germanium, which acts as a donor on the 3-valent lattice
site. Consequently the desired highly doped layer is formed
between the metal and the bulk semiconductor.

The optimum alloying temperature has to be found for each
alloy-semiconductor combination. In Table 3-5 metals and
optimized alloying or sintering temperatures are collected
for contacts important in LED technology.

P type GaAs, GaP and the mixed compounds between these can be
provided with ohmic contacts by Au:Zn. Zinc acts not only as
an acceptor in the dissolved and then regrown semiconductor,
but moves also by diffusion into the bulk material.

Aluminum is a very favorable contact metal for p type GaAs
and the mixed compounds $GaAs_{1-x}P_x$ up to x = 0.85. During heat
treatment the aluminum does not melt, the optimum time of
sintering is therefore much longer than for alloyed contacts.
For GaAs low resisitivity contacts are obtained by sintering
at 480° C for 45 min. Another example of a non-alloyed contact is a three layer film of Ti/Pt/Au, which can be used on
p-type GaAs.

Table 3-5

Contacts of technological importance for III-V compounds

| III-V | Type of conductivity | Contact material | Heat treatment |
|-------|----------------------|------------------|----------------|
| GaAlAs | n | Au:Ge | 450° C/5 min. |
| GaAs | n | (1 to 12 % Ge) | |
| GaAs | n | Au:Sn (10 % Sn) | 450° C/5 min. |
| GaAsP | n | Au:Ge (1 to 12 % Ge) | 480° C/5 min. |
| | | Au:Sn (10 % Sn) | 480° C/5 min. |
| GaP | n | Au:Ge (1 to 12 % Ge) | 580° C/5 min. |
| | | Au:Sn (10 % Sn) | 600° C/5 min. |
| InP | n | Au:Sn (10 % Sn) | 420° C/5 min. |
| GaAlAs | p | Au:Zn (1 to 3 % Zn) | 500° C/5 min. |
| | | Al | 600° C/5 min. |
| GaAs | p | Au:Zn (1 to 3 % Zn) | 500° C/5 min. |
| | | Al | 480° C/45 min. |
| | | Ti/Pt/Au | deposited at 150° C |
| GaAsP | p | Au:Zn (1 to 3 % Zn) | 500° C/5 min. |
| | | Al | 480° C/45 min. |
| GaP | p | Au:Zn (1 to 3 % Zn) | 600° C/5 min. |
| InP | p | Au:Zn (1 to 3 % Zn) | 450° C/5 min. |
| | | In:Zn (5 % Zn) | 400° C/5 min. |

The two latter contacts have a big advantage compared with
all gold based alloyed contacts: they are easily wire bonded.
It is very difficult to achieve good bond adhesiveness on
alloyed gold. The reason is probably a layer of oxidized gallium on top of the gold metallization as well as the fact
that all relevant gold alloys are extremely hard. To facilitate wire bonding a blocking layer of Ti:W (10:90), Ta or W
can be inserted between the contact and a cover of pure gold.

### 3-3-5 Chemical etching

Hundreds of etches are used in semiconductor technology for a
variety of aims. To offer some insight into etches and applications in limited space, Table 3-6 contains a selection of
etches which are useful for LED production. In the text only
basic principles of wet chemical etching are described and
examples used where thought to be useful.

Pattern definition is one of the main applications of etch
processes. Depending on the material into which a structure
has to be etched, the dissolution of a metal, a dielectric or
the III-V compound itself is necessary. The etch process
should act selectively with respect to the resist and the
material to be etched. Further aspects are reasonable etching
time at reasonable temperature.

Other applications of etches are removal of handling damage,
especially from sawing and grinding. Cleaning procedures
based on etches are used prior to sensitive process steps
like epitaxial growth and diffusion. Finally etches serve as
analytical tools to make pn junctions and epitaxial layers
perceptible.

From a practical point of view one of the most important
properties of an etch is the etch rate, usually measured in

µm/min. Certainly the etch rate is an especially useful parameter, if it is independent of etching time. However, this
independency is not always given. In an etching process particles, usually ions, move through a solution to the solid
surface where a chemical reaction occurs. The products of the
reaction, which should be soluble in semiconductor etches,
are dispersed again in the solution. Without stirring or
other movements the mass transport in the solution is accomplished solely by diffusion. The overall speed of the reaction and therefore the etch rate can be limited essentially
by two mechanisms: either by the transport speed of the chemical species in the solution or by the chemical reaction at
the solid surface. In the first case the process is therefore
called diffusion limited, in the second case reaction limited. From these two limitations rather different properties of
the etching process may be deduced.

Diffusion limited reactions remove the material proportional
to the square root of the etching time, a consequence of the
diffusion laws explained in section 3-3-3. Therefore in this
case there is no constant etch rate. Moreover the etch rate
is strongly dependent on stirring or movements of the wafers.
Both limiting conditions, violent stirring or pure diffusion,
are difficult to realize, because diffusion is easily im-
paired by unintended thermal convection. Therefore in all
cases where etching depth is of prime importance diffusion
limited etches should be avoided.

Diffusion limited processes have usually lower activation
energies than reaction rate controlled processes. Consequently the diffusion limited processes are less temperature dependent than reaction limited processes. Therefore an etch
may be reaction rate limited at low temperature but diffusion
limited at high temperature. Similarly the limitation may depend on the concentration or composition of the etch and even

Table 3-6

Etchants used for III-V wafer processing

| Etchants for GaAs | Etch rate µm / min | Remarks |
|---|---|---|
| $H_2SO_4$:$H_2O_2$:$H_2O$, 8:1:1, 25° C | 1.0 | specular |
| $H_2SO_4$:$H_2O_2$:$H_2O$, 4:4:2, 25° C | 6.0 | rough |
| $H_3PO_4$:$H_2O_2$:$H_2O$, 2:6:2, 30° C | 4.0 | anisotropic |
| $CH_3OH$:$Br_2$, 99:1, 25° C | | polishing or anisotropic depending on orientation |
| $H_2O_2$:NaOCl, 20:1, 25° C | 0.15 | polishing with pad |
| $NH_4OH$:$H_2O_2$:$H_2O$, 2:1:5, 25° C | 2.5 | mesa etch |
| $H_2O_2$:$NH_4OH$, 10:1, 25° C | 6 | high etch rate for GaAs, low for GaAlAs |
| HF:$H_2O_2$, 1:1, 25° C, light | - | pn junction |

| Etchants for GaP | Etch rate µm / min | Remarks |
|---|---|---|
| $H_2SO_4$:$H_2O_2$:$H_2O$, 8:1:1, 60° C | 0.1 | cleaning |
| $H_2SO_4$:$H_2O_2$:$H_2O$, 3:1:1, 60° C | 0.2 | |
| $K_3Fe(CN)_6$:KOH, 1 Mol: 0.5 Mol, 60° C | 2.0 | mesa etch |
| HF:$H_2O_2$, 1:1, 25° C, light | - | pn junction |

...

Table 3-6 (continued)

| Etchants for different materials | Material | Etch rate µm / min | Remarks |
|---|---|---|---|
| $CH_3OH:Br_2$, 99:1, 25° C | InP | 1.5 | polishing |
| $NH_4F:HF$, 6:1, 25° C | $SiO_2$ | 1.5 | pattern definition |
| $H_3PO_4$, 160° C | $Si_3N_4$ | 0.03 | pattern definition |
| $NH_4F:HF$, 6:1, 25° C | $Al_2O_3$ | 0.5 | pattern definition |
| $H_3PO_4$, 60° C | Al | 0.5 | pattern defintion |
| $KI:I_2:H_2O$ 4g:1g:40g, 25° C | Au | 0.5 | pattern definition |
| $K_3Fe(CN)_6:KCN:KOH$ 0.4Mol:0.2Mol:0.1Mol/ 25° C | Au | 0.06 | pattern definition |

the crystal orientation of the semiconductor surface which
has to be etched. The etch rate depends nearly always to some
extent on the orientation of the surface of a single crystal.
If the dependency is not very pronounced the etch is called
isotropic, otherwise anisotropic. For polishing purposes any
anisotropic reaction should be avoided, therefore the purely
chemical reaction is usually combined with mechanical abra-
sive action in a chemo-mechanical process.

Many etch reactions, especially of semiconductors, involve an
oxidation-reduction process followed by the dissolution of
the oxidation products. Typical examples of such etch systems
are mixtures of $H_2SO_4 : H_2O_2 : H_2O$, or $H_3PO_3 : H_2O_2 : H_2O$, or

$NH_4OH$ : $H_2O_2$ : $H_2O$. Etch rates are displayed most appropriate
in ternary diagrams as shown in Fig. 3-27 for the system
$H_3PO_4$ : $H_2O_2$ : $H_2O$ for GaAs according to results by MORI and
WATANABE (1978). In this figure constant etch rate contours
are shown over a triangular area in which every point corresponds to a certain composition of the etch. At 30° C the
highest etch rate of about 4.5 µm/min is obtained at a composition of 30 % $H_3PO_4$ and 70 % $H_2O_2$. The pure constituents of

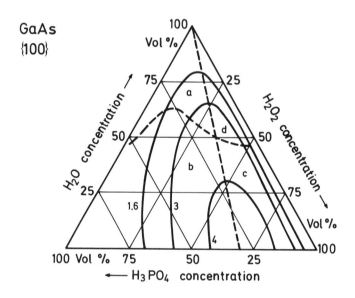

Fig. 3-27   The etch system $H_3PO_4$ : $H_2O_2$ : $H_2O$, for
GaAs at 30° C

the etch system do not etch GaAs, neither will binary mixtures between $H_2O$ and $H_3PO_4$ or $H_2O$ and $H_2O_2$. The diagram is
divided by the broken lines into four parts a, b, c and d,
which are characterized by different rate limiting processes.
Etches whose compositions fall into regions a, c and d operate reaction limited, compositions within region b are
limited by $H_2O_2$ diffusion.

Specific examples of etches and their application are collected in Table 3-6.

## 3-3-6 Probing

Wafer probers are instruments which are used to test the quality and yield of completely processed wafers. The prober
puts a probe on the contact of a first diode on the edge of a
wafer, allowing the user to perform measurements with his own
equipment as suited for his purposes. After the measurements
on the first diode have been finished, the prober moves the
wafer a controlled distance to the next diode. The data obtained during a measurement cycle are either immediately evaluated in an accept-reject manner and bad diodes marked by a
dot of ink, or the data are stored together with a diode's
position and are later evaluated.

The prober communicates with the controlling computer in a
handshake type manner: a signal from the prober tells the
computer: probe on the device, start test. After measurements
have been finished and the inker has been eventually activated, the computer tells the prober: test done, move on. When
the edge of a wafer is reached, a mechanical or optoelectronic edge sensor causes the prober to turn to the next row of
diodes and allows the computer by a special signal to keep
track of the localization of any diode.

The test equipment used for LED testing is usually selected
to measure optical properties like light flux or radiation
flux and electrical data like forward voltage and reverse
current or breakdown voltage. The cycle time for measurements
can be shortened by using more than one, frequently four,
contact probes. Typical cycle times range between 0.2 and 0.6
seconds for one LED.

## 3-3-7 Dicing

Dicing is a process that is much more important for LED production than might be expected. By creating new reflecting
surfaces, dicing can improve the light flux of diodes on
transparent substrates by a factor of 2.0 to 2.5 if it is
done properly. On the other hand it can reduce the device
lifetime severely when done incorrectly.

There are essentially two techniques to produce dice from a
III-V wafer: sawing and scribing. For scribing usually diamond scribe machines are preferred over laser scribers to ex-
clude health risks by the vaporizing group V component. The
scribed wafers have to be broken into dice by controlled
bending of the wafer for example by a cylindrical roller.
During breaking wafer and dice are held on an adhesive foil.
This foil is afterwards used to separate the freshly produced
dice by stretching. The separation is necessary to facilitate
the picking of a die by a collet which transports the die to
the mounting station.

The scribe technique has the big advantages that the gain in
light output is there immediately after breaking and is very
reproducible and that lifetime problems are never encountered. On the other hand, there are drawbacks. The stretching
of the foil does not move the dice exactly the same distance
apart. Much worse, two or more dice may cling together, if
the metallization of the dice is not everywhere completely
ruptured. This is very disadvantageous for automatic mounting
techniques. Also, and very important, the first mask in wafer
processing has to be aligned parallel to the ⟨110⟩ directions
of the crystal, otherwise scribed wafers will not break accu-
rately. That means that only $\{100\}$ oriented wafers can be
processed into rectangular dice by scribing and breaking.
$\{111\}$ oriented substrates, frequently used for liquid phase

epitaxy, are not suited for scribing because they can be broken only into triangular pellets.

The second dicing method is sawing. For this purpose saws
with circular diamond reinforced blades are used. The wafer
is held on an adhesive foil as for breaking of scribed wafers. Stretching is not necessary because the dice are separated by at least the equivalent of the width of the blade,
typically 30 μm. Therefore all dice are kept in excellent
order. Moreover all orientations of wafers can be handled
equally well. However, there are disadvantages of this technique too. The sawn surface is heavily damaged. The damage
manifests itself in strain that can be made visible between
crossed polarizers. Strained diodes degrade severely during
life-testing. But the damage also prevents the expected increase of light flux. For these two reasons the sawing damage
has to be removed. This can be done by chemical etching of
the sawn wafer. An etching depth of 0.2 to 0.5 μm is sufficient. Suitable etches for GaAs are a mixture of $H_2SO_4$ :
$H_2O_2$ : $H_2O$ in the ratio 5 : 3 : 2 and for GaP a solution of
0.5 mole KOH and 1 mole $K_3Fe(CN)_6$ in 1 liter of $H_2O$.

## Questions

Q3-1    What amounts of gallium and arsenic are required to
        grow a GaAs crystal of 1 kg by the HB method, if one
        assumes that the vapor of arsenic consists of $As_4$
        molecules and exhibits ideal gas behavior, and that
        the volume of the ampoule used is 2 l, at a tempera-
        ture of 610 °C?

Q3-2    Derive a simple formula which gives an estimate of the
        maximum allowed cooling rate in LPE, assuming diffu-
        sion transport with diffusion constant D, maximum
        supercooling ΔT, and solution thickness t.

Q3-3    What is the thickness of a GaP epitaxial layer, grown
        from a 1 mm thick melt saturated at 1000 °C (solubi-
        lity of P in Ga 1.8 at %), assuming complete deposi-
        tion?
        (Densities of Ga and GaP: 5.9 and 4.13 $g/cm^3$, molecu-
        lar weights 69.7 and 100.7, respectively).

Q3-4    Calculate the capacitance of a square shaped bonding
        pad 100 x 100 $\mu m^2$ on a $Si_3N_4$ layer 100 nm thick
        ( $\varepsilon_0$ = 8.85 x $10^{-12}$ As/Vm, $\varepsilon$ see Table 3-2).

Q3-5    Calculate the temperature dependence of the diffusion
        constant of Zn in GaAs using the experimental results
        displayed in Fig. 3-23. Assume a $Zn_3As_2$ diffusion
        source, diffusion temperatures of 730 °C and 650 °C
        and ratios between background and surface concentra-
        tions of $N_b/N_0$ = $10^{-3}$ and $10^{-4}$.

## References

ANDRÉ, J.P., A. GALLAIS and J. HALLAIS, "GaAs-GaAlAs hetero-
    structures grown by the metal-alkyl process," Inst.
    Phys. Conf. Ser., 33a (1977), pp. 1-8

BASS, S.J. and P.E. OLIVER, "Pulling of gallium phosphide
    crystal by liquid encapsulation," J. Crystal Growth 3,4
    (1968), pp. 286-290

BLUM, J.M. and K.K. SHIH "Growth of smooth uniform epitaxial
    layers by liquid-phase-epitaxial method" J. Appl. Phys.,
    43 (1972), pp. 1394-1396

CASEY Jr., H.C. and M.B. PANISH, "Reproducible diffusion of
    Zn into GaAs: application of the ternary phase diagram
    and the diffusion and solubility analyses", Trans. of
    the Metall. Soc. of AIME, 242 (1968), pp. 406-412

CRAFORD, M.G. and J.J. HOPFIELD, "Vapor phase epitaxial ma-
    terials for LED applications," Proc. IEEE, 61 (1973),
    pp. 862-880

GILLESSEN, K., A.J. MARSHALL and J. HESSE, "Temperature gra-
    dient solution growth, application to III-V semiconduc-
    tors," Crystals, 3, Springer: Berlin (1980), pp. 49-71

KÖHL, F. and F.G. VIEWEG-GUTBERLET, "Gallium-arsenide, the
    material and its application," Microelectronics Journal,
    12 (1981), pp. 5-8

MORI, Y. and N. WATANABE, "A new etching solution system,
    $H_3PO_4$ : $H_2O_2$ : $H_2O$ for GaAs and its kinetics," J. Elec-
    trochem. Soc., 125 (1978), pp. 1510-1514

PLOOG, K., "Molecular beam epitaxy of III-V compounds," Cry-
    stals, Vol. 3, Springer: Berlin (1980), pp. 73-162

SAUL, R.H. and O.G. LORIMOR, "Liquid phase epitaxy processes
    for GaP LED's," J. Crystal Growth, 27 (1974), pp.
    183-192

SHIH, K.K., "High surface concentration Zn diffusion in
    GaAs", J. Electrochem. Soc., 123 (1976), pp. 1737-1740

STRINGFELLOW, G.B., "Thermodynamic aspects of organometallic
    vapor phase epitaxy", J. Crystal Growth, 62 (1983),
    pp. 225-229

STRINGFELLOW, G.B., M.E. WEINER and R.A. BURMEISTER, "Growth
    and properties of VPE GaP for green LEDs," J. Electronic
    Materials, 4 (1975), pp. 363-387

THURMOND, C.D., "Phase equilibria in the GaAs and the GaP
    systems," J. Phys. Chem. Solids, 26 (1965), pp. 785-802

TUCK, B. and P.R. JAY, "A radiotracer study of Zn diffusion
    profiles of GaP at 900° C," J. Phys. D. Appl. Phys., 10
    (1977), pp. 1315-1322

# 4 TYPES OF LED

We will describe the structures of different LED chips in this chapter. Again we will concentrate on the more common types, which are produced on an industrial scale, or which are at least announced to be on the verge of mass production. We begin with visible LEDs of all colors from red to blue and treat infrared emitters of different kinds in the second place.

## 4-1 Visible LEDs

### 4-1-1 Red Leds

There are three different types of red LED available: standard red (or GaAsP red), GaP red, and GaAlAs red (which is also sometimes named super-red, extra-bright-red or similar, depending on the manufacturer). Typical cross-sections of these chip types are shown in Fig. 4-1.

The active region of a standard red LED (see Fig. 4-1, left hand side) consists of $GaAs_{1-x}P_x$ with a Phosphorus content x of about 40 %, corresponding to an emission wavelength of about 660 nm. This layer is deposited by a vapor phase epitaxial technique on an gallium arsenide substrate with a transition layer of varying composition in between (see section 3-2-2). The epitaxial layer has a typical total thick-

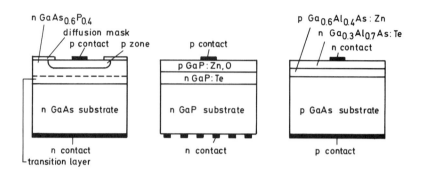

Fig. 4-1 Cross sections of red LED chips, left:
        standard red (GaAsP), center: GaP:Zn,O red,
        right: high intensitiy red (GaAlAs)

ness of 50 μm and is doped n type in the $10^{17}$ $cm^{-3}$ range,
mostly using tellurium as dopant. The active pn junction is
produced in a planar type of wafer process by masked zinc
diffusion with a diffusion depth of about 5 μm and a surface
concentration typically around p = $10^{19}$ $cm^{-3}$. The p side contact consists of aluminum or a gold-zinc alloy and covers
only a small fraction of the surface of the chip, to allow
unobstructed light emission from the pn junction. Light emitted towards the n side is virtually completely absorbed by
the GaAs substrate, because the substrate thickness is typically 200 μm, and the absorption length of red light with a
wavelength of 660 nm in GaAs is only about 1 μm. The n side
is contacted with an alloyed layer of e.g. Au-Ge. A scanning
electron microscope photograph of a standard red LED chip is
shown in Fig. 4-2.

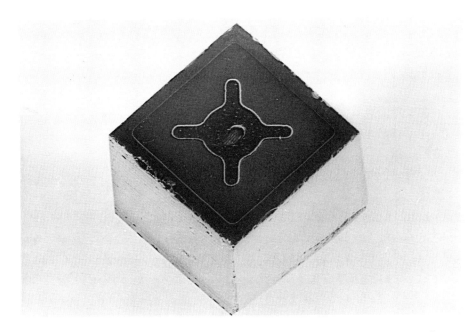

Fig. 4.2 Standard red LED chip (SEM photograph),
        showing cross shaped p side contact with
        round bonding area in the center. An impres-
        sion from the probing tip can be seen in the
        center. The square shaped diffusion window
        is also visible on the top surface. The side
        faces of the chip are sawn. The chip side
        length is about 240 μm.

The main factors influencing the brightness of this type of
LED are the materials quality of the $GaAs_{0.6}P_{0.4}$ layer and
the depth of the Zn diffusion. If the diffusion is too shallow, injected electrons diffuse to the chip surface where
they recombine nonradiatively. If the pn junction is too
deep, a considerable part of the light is reabsorbed along
its path to the surface. The optimum depth is in the range of
two to three times the diffusion length of the injected electrons. Therefore, a diffusion depth of around 5 μm is chosen.

This value together with the high hole concentration is also sufficient to provide a sufficiently low sheet resistance for efficient current spreading from the small area p contact to the relatively large area pn junction.

Standard red LEDs of the kind just described are being produced in very large quantities. Their main advantage is the low price which has opened a very wide field of applications. Their brightness is however surpassed by other types of LEDs.

The second type of red LED consists completely of gallium phosphide GaP (see Fig. 4-1, center). Red light with a wavelength of 690 nm is produced by recombination via the isoelectronic center Zn-O in the p layer (see section 2-2). The structure is fabricated by double liquid phase epitaxy, the first layer being doped with tellurium, and the second layer with zinc and oxygen. This simultaneous doping provides both Zn acceptors for p type conduction and Zn-O pairs as isoelectronic recombination centers. In contrast to standard red LEDs, the emitted light is hardly absorbed in the epitaxial layers or in the substrate, because the band gap of GaP is considerably higher than the photon energy. Therefore, the backside contact is only alloyed to a small fraction of the substrate, leaving the main part of the reverse side reflecting. This results in a nearly twofold increase in brightness.

Red LEDs on the basis of GaP:Zn,O exhibit quite high quantum efficiencies, because the Zn-O center leads to very efficient emission, and because reabsorption in GaP in negligible. In fact, the world record in efficiency for visible LEDs was achieved with GaP:ZnO diodes: it is 15 %. Nevertheless these LEDs play only a minor role from two reasons: first, the emission of 690 nm is already in a spectral region where the sensitivity of the human eye is rather low, so that the high efficiency does not translate into considerably higher

brightness than standard red, and second, the high efficiency
is attained only at low currents; it decreases with higher
currents, resulting in a sublinear brightness-current dependence. This latter property is especially disadvantageous for
multiplex operation of LEDs, where the devices are driven
with short pulses of increased current to achieve the same
brightness as with continuous operation. Consequently, the
use of GaP red LEDs is diminishing.

The opposite is true for the third type of red LEDs, the
GaAlAs very bright red LEDs. An example of this device is
shown in Fig. 4-1 on the right hand side. This structure is
fabricated by double liquid phase epitaxy of $Ga_{1-x}Al_xAs$ on
p-type GaAs substrates. The first layer, which is also p-type
by Zn doping beyond $10^{18}$ $cm^{-3}$, has an aluminum content of
40 % corresponding to an emission wavelength of 650 nm and
forms the recombination region of the device. The second
layer with an aluminum content of 70 % and n-doping serves as
an efficient electron injector (the two epitaxial layers form
a heterojunction), and at the same time as ·a transparent
window for the emanating radiation. Because the substrate
absorbs red light, the backside contact is again provided on
the full area, as in the case of standard red.

These GaAlAs red LEDs can be up to ten times brighter than
standard red. The reason for this superior performance is
fourfold: first, the injection efficiency is practically
100 % due to the heterojunction; second, the quality of the
material of LPE layers is much better than with VPE; third,
the radiation efficiency in p type material is higher than in
n type material, and fourth, there is practically no reabsorption in the high band gap top layer. Due to their considerably higher brightness, GaAlAs red LEDs are used in rapidly increasing numbers and are expectecd to replace standard
types to some extent, as soon as prices fall as a consequence

of the increasing production volume.

Even with these very bright GaAlAs LEDs, half of the light
generated at the pn junction is lost by absorption in the
GaAs substrate. Therefore, another factor of two in brightness can be gained if the the absorbing GaAs substrate is
replaced by a transparent crystal. This idea was realized
recently (see ISHIGURO et al. 1983), but the required fabrication sequence is rather complicated: in a first epitaxial
process a thick (about 200 μm) layer of GaAlAs with an aluminum content of more than 40 % is grown on a GaAs substrate,
which is subsequently removed by selective chemical etching.
This epitaxial layer then serves as a substrate for the light
generating heterostructure, which is grown in a second epitaxial process. Because the active layer of this structure has
the lowest bandgap, there is no reabsorption in the surrounding layers. It remains to be seen whether this complicated
device can be produced economically.

## 4-1-2 Orange and yellow LEDs

Orange and yellow LEDs are dealt with jointly because they
are fabricated using a common technology. Cross sections of
orange (620 nm) and yellow (590 nm) emitting chips are depicted in Fig. 4-3.

Basically, these LEDs are very similar to the standard red
LEDs described in chapter 4-1-1: they consist of an active
layer of $GaAs_{1-x}P_x$ which is deposited using a vapor phase
epitaxial method. The pn junction is produced by masked Zn-
diffusion into the active layer. Of course, there are some
differences to red LEDs: for orange emission the composition
is adjusted with a phosphorus content of 60 %, and for yellow
with 85 %. Because both compositions are in the indirect
range, nitrogen is added as an isoelectric center to in-

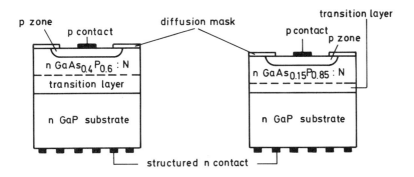

Fig. 4-3 Cross sections of vapor phase epitaxial planar $GaAs_{1-x}P_x$ LEDs, left: orange with x =
0.6, right: yellow with x = 0.85

crease the radiative recombination efficiency (see section
2-3 and 6-1-2). The lattice constants of the materials for
orange and yellow LEDs are closer to that of gallium phosphide than that of gallium arsenide. Therefore GaP is used as
a substrate here. This gives the additional advantage that
the substrate is transparent to the emitted light. As can be
seen from Fig. 4-3, structured substrate contacts to achieve
a reflecting reverse side are provided to increase the light
output. In the transition layer between substrate and active
layer the lattice constant is gradually changed. The corresponding change of composition must be kept below a rate of
approximately 1 %/1 μm to achieve the required quality of
material in the active layer. As a consequence, the transition layer in orange LED material must be at least 40 μm
thick, because it has to bridge a composition difference of
40 %. The total epitaxial layer thickness is therefore typically around 80 μm. For yellow emitting material the transition layer can be accordingly thinner (composition difference

only 15 %), which is also indicated in Fig. 4-3.

Although the material for orange and yellow material is in
the indirect range, the brightness of these LEDs can be higher than that of standard red. The explanation for this possibly surprising fact is that the sensitivity of the eye increases very much with decreasing wavelength in this spectral
region (see section 5-1), and that there is no absorption
loss in the substrate. LEDs with an emission wavelength of
about 630 nm (reddish-orange) are named "high efficiency red"
by some manufacturers. We prefer however to classify these
devices under "orange", because of the GaP based technology.

Other materials for orange and yellow LEDs have also been
considered. For example, indium gallium phosphide $In_{1-x}Ga_xP$
could in principle be used for light emission in this spectral range (see also section 2-4). The technology of this
material is however much more difficult than that of
$GaAs_{1-x}P_x$, so that InGaP LEDs have only been realized on an
experimental scale. These devices did not come up to the
brightness of GaAsP LEDs. Yellow LEDs were also fabricated
using highly nitrogen-doped GaP as the active material. This
technological approach was abandoned, because there is no
particular advantage over the existing technology. Yellow
devices based on silicon carbide (SiC) were the first LEDs on
the market; their brightness was rather low, however, so that
they did not survive.

### 4-1-3 Green LEDs

Green LEDs are made from gallium phosphide GaP, either with
vapor phase or liquid phase epitaxy. Examples of both types
of devices are shown in Fig. 4-4. The vapor phase epitaxial
GaP LED in the left part of the figure exhibits a structure
analogous to the orange and yellow LEDs of section 4-1-2. A

nitrogen-doped epitaxial layer is deposited by vapor phase epitaxy on a GaP substrate. The p region is fabricated by masked Zn-diffusion into the surface of the n conducting epitaxial layer. A small area p contact and a reflecting n contact are provided as with the other LED types with a transparent substrate. The liquid phase epitaxial green LED as shown in the right part of Fig. 4-4 consists of two epitaxial layers (n and p type, respectively) which are grown on a GaP substrate either with a two melt type or more frequently with a one-melt liquid phase epitaxy process with doping via the vapor phase (see section 3-2-1). The example shown here is a mesa type of chip, which can be tested before the semiconductor is diced. A scanning electron microscope photograph of a chip of this type is shown in Fig. 4-5.

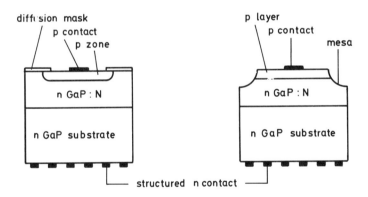

Fig. 4-4 Cross sections of green GaP LEDs, left:
        vapor phase epitaxial planar type, right:
        liquid phase epitaxial mesa type

Fig. 4-5 Mesa type GaP green LED chip (SEM photo-
         graph), showing cross shaped p side contact.
         The mesa measures about 240 μm by 240 μm,
         whereas the chip side length is about
         310 μm.

Although vapor phase epitaxial green LEDs can be produced
very economically with basically the same technology as red
(standard), orange, and yellow diodes, they are more or less
superseded by their much brighter liquid phase epitaxial com-
petitors. The LPE process for production of green GaP LEDs
has been rationalized very much recently. Therefore, the economic advantage of VPE green is only marginal now.

Brightness and color of GaP green LEDs are strongly influenced by the nitrogen content around the pn junction. With increasing nitrogen concentration the brightness tends to increase (because of the increasing number of recombination

centers available), and the emission color is shifted into
the yellow region of the spectrum, because a larger fraction
of the injected carriers recombines with slightly lower energy at nitrogen pairs, which are formed with higher probability at larger nitrogen concentration (see section 6-1-2-1).
The second factor influencing the efficiency of GaP LEDs is
the minority carrier lifetime in the n region, where most of
the light is generated. The minority carrier lifetime in turn
depends on the doping level (n concentrations as low as possible are aimed at), on the purity, and on the number of
structural defects. With epitaxial layers of extremely high
quality nominally nitrogen-free GaP diodes are fabricated,
which emit pure green light with a dominant wavelength of
about 555 nm, with reasonable efficiency. Most green LEDs
which are available today however contain nitrogen in concentration in excess of $10^{18}$ cm$^{-3}$. Their dominant wavelengths
are mainly in the region from 565 to 570 nm, corresponding to
a more or less yellowish green emission color. Nitrogen-doped
green LEDs are generally brighter than nitrogen-free diodes.

Although there is no lattice matching problem in the epitaxial growth of GaP on GaP, the n layers of LPE green diodes
are usually grown at least 50 μm thick to achieve low doping
and high materials quality at the pn junction. The p layer
thickness is mostly in the range 10 to 20 μm and is less critical than in the case of standard red diodes. To minimize
reabsorption, nitrogen is only incorporated in the region
where the recombination takes place, i.e. in the vicinity of
the pn junction up to a distance of 2-3 diffusion lengths,
which amounts to around 15 μm.

The brightness of GaP green LEDs is especially dependent on
the quality of the material. For example, dislocations lead
to dark areas in LEDs which have a radius approximately equal
to one diffusion length. As dislocations are transferred from

the substrate to the epitaxial layer during liquid phase epitaxy, the substrate itself should have the lowest possible
dislocation density. With a typical dislocation density of
$10^5$ $cm^{-2}$, an LED with an area of 0.1 $mm^2$ contains a number of
100 dislocations. If the diffusion length is assumed to be
7 $\mu m$, 100 dislocations affect about 1.5 x $10^{-2}$ $mm^2$ or 15 % of
the LED surface area. This consideration shows that the dislocation density is negligible only if it is kept definitely
below $10^5$ $cm^{-2}$.

## 4-1-4 Blue LEDs

Because blue light corresponds to photon energies in excess
of 2.5 eV, LEDs for this color have to be fabricated from a
completely different class of semiconductors. The technology
of these materials is far less advanced than that of the GaAs
and GaP based compounds. It is therefore not surprising that
blue LEDs cannot compete with the common red to green ones
with respect to brightness or efficiency. Nevertheless blue
LEDs are available today, consisting either of gallium nitride (GaN) or silicon carbide (SiC), as shown in Fig. 4-6.

The GaN device (left side of Fig. 4-6) is shown in the position as it is mounted, with the substrate up and the active
layer down. GaN cannot be produced in bulk single crystalline
form, because the dissociation pressure at the yet unknown
melting point of this compound would be extremely high (probably larger than $10^5$ bar). Single crystalline layers can
however be grown epitaxially on sapphire substrates using a
variation of the vapor phase chloride process which is described in section 3-2-2. Undoped GaN is usually highly n conducting with a typical carrier concentration around
$10^{19}$ $cm^{-3}$, presumably due to nitrogen vacancies. These native
donors can be compensated by heavy doping with zinc acceptors, so that insulating material results. It is not possible

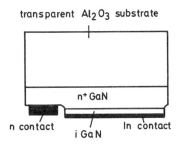 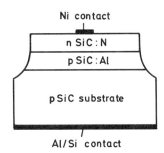

Fig. 4-6 Cross sections of blue LED chips, left: GaN
device on Al$_2$O$_3$ substrate, right: double
liquid phase epitaxial SiC device

up to now to achieve p type behaviour, at least with practicable high conductivity. For fabrication of GaN blue LEDs an
undoped n$^+$ layer is first grown on a sapphire substrate, followed by a very thin (about 0.1 μm) Zn doped insulating
layer. After removal of part of the i layer, metallic contacts are deposited on the n$^+$ and i layer, respectively. If a
voltage of typically 5 to 10 V is applied to the contacts
with the i contact positive, a high electric field region
exists in the i layer. Electrons from the n$^+$ region are injected into the i region where they can recombine radiatively
at the Zn acceptors. Blue light is emitted through the transparent Al$_2$O$_3$ substrate. Because these devices do not contain
a pn junction, it is formally not correct to use the designa-
tion light emitting diodes. Nevertheless it is common practice to name them LEDs.

Besides their limited brightness GaN blue LEDs have two further disadvantages: the forward voltages are rather high
(typically 5 to 10 V), and the maximum allowed current is low
(about 10 mA), which makes them less suited for pulsed multiplex operation.

A typical blue LED chip made from SiC is shown in the right
part of Fig. 4-6. It consists of an SiC substrate which is
grown by a kind of sublimation process at temperatures around
2500° C. This technique yields only small platelets of up to
10 mm side length, which must be machined with diamond tools
because of the extreme hardness of SiC. Two layers of SiC are
then deposited on the substrate using a liquid phase epitaxial process at about 1800° C from a silicon rich solution.
SiC can be doped n or p type with nitrogen or aluminum, respectively. Ohmic contacts to SiC consist for example of Ni (n
side) and Al/Si alloy (p side). The mesa etching process indicated in Fig. 4-6 is carried out at about 1000° C in an
oxygen and chlorine containing atmosphere.

Generally, all technological steps required for LED fabrication from SiC are possible and developed to some extent. They
are, however, much more difficult to perform than with other
semiconductors, due to the very high temperatures involved
and to the extreme mechanical and chemical stability of silicon carbide. As it was stated once: "Silicon carbide is very
much like silicon, only harder."

The efficiency of SiC LEDs is rather low compared with the
GaAs and GaP based types, because SiC is an indirect semiconductor. No efficiency enhancing isoelectronic trap analogous
to nitrogen in GaP is known, so that radiative recombination
can only occur via donors and/or acceptors. Therefore, the
prospects of improving the brightness of SiC LEDs are rather
bad.

To summarize, blue LEDs with moderate brightness can be fabricated from GaN or SiC. To make blue LEDs really competi-
tive with the common red to green types, a basic improvement
is required, however, which could be a new material or a yet
unknown trick in the technology of the old materials.

## 4-2 Infrared emitters

Infrared emitting diodes, or IREDs, are defined throughout
this book as those emitting in the near infrared spectral
region between the visible and about 950 nm. The generally
observed tendency to replace vapor phase epitaxy by liquid
phase epitaxial techniques has led to the abolition of all
vapor phase methods for the production of IREDs. The reason
is the always higher optoelectronic efficiency of the liquid
phase epitaxial material, which is probably due to a reduced
Ga vacancy concentration compared with vapor phase material.

### 4-2-1 Standard types

By standard types we understand the mass produced Si doped
GaAs diodes and the newer GaAlAs diodes, also with Si doping.
Silicon offers some extremely favorable features as a doping
material. First, as an element of the fourth column of the
periodic table, it can occupy either Ga or As sites of the
GaAs lattice. Correspondingly it can act either as a donor or
as an acceptor, making pn junctions possible with the same
doping species on both sides of the junction. Second it is
very easy to control the doping behavior of silicon simply by
controlling the temperature of the liquid epitaxial growth.
On $\{111\}$ B faces (cf. Fig. 1-3) a layer grows n type at temperatures above about 820° C and p type below. Because growth
in liquid phase epitaxy is usually induced by controlled lowering of the temperature, n and p type layers can be grown
in a quite natural way. Third, the optical emission caused by

silicon doping is shifted well below the band gap energy.
This is due to the formation of band tails as explained in
chapter 6. Therefore the reabsorption of radiation produced
close to the junction is reduced and the external efficiency
improved.

The amphoteric Si doping, described above, works in GaAs as
in GaAlAs. Thus the emission wavelength can be tailored to
special applications by adjusting the composition of the mixed crystal GaAlAs.

Two typical structures of GaAs:Si and GaAlAs:Si diodes are
displayed in Fig. 4-7. The epitaxy of both types is performed
by a single melt producing the n as well as the p zone. In
the case of the left diode, the GaAs:Si type, the epitaxial
layers are grown on an n type GaAs substrate. The doping
level of the substrate should be below $1 \times 10^{18}$ $cm^{-3}$ to provide for a sufficiently low free carrier absorption around
the peak emission wavelength of 940 to 950 nm. On the other
hand it should be above $1 \times 10^{17}$ $cm^{-3}$ to offer an acceptable
series resistance in pulse applications typically reaching up
to 2.5 A. The epitaxy has to be performed around 820° C, the
temperature of the pn junction formation. Above 820° C the n
zone with a typical thickness of 50 μm is grown; below that
temperature the p zone grows with a thickness of typically
40 μm. After completion of the epitaxy, the rather simple
steps of wafer processing are started. At first the p side
contact is made of Al or Au:Zn. After that the p side is
covered by a suitable material, e.g. $SiO_2$, which has to serve
as a mask for the mesa etch done in a $NH_4OH:H_2O_2:H_2O$ solution. The etching depth has to be larger than the p zone
thickness. After removal of the mask material the p side is
complete. The reverse side is thinned to about 250 μm and
metallized by Au:Ge and alloyed. The alloyed metal has a very
low optical reflectivity and therefore absorbs radiation im-

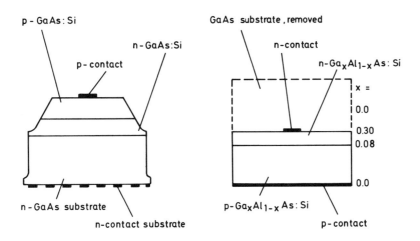

Fig. 4-7 Cross sections of standard IREDs, left
GaAs:Si diode, right GaAlAs:Si diode

pinging on it. It is therefore advantageous to take special
measures to cover only a small part of the n side by the
alloyed contact material and to ensure a high optical reflectivity of the material between alloyed regions, e.g. by etching in a polishing etch. The wafer can then be probed and
diced. $\{111\}$-oriented wafers, which are usually preferred,
have to be sawn and should be etched afterwards to remove
sawing damage and to further improve the efficiency. A scanning electron microscope (SEM) photograph of a GaAs:Si diode
is shown in Fig. 4-8.

The simple structure of the GaAlAs:Si diode, shown on the
right side of Fig. 4-7, is rather difficult to produce. To
understand this, one has to consider the Al profile inside
the diode.

Fig. 4-8 Mesa type GaAs:Si chip (SEM photograph),
exhibiting a finely structured p side contact for improved current distribution during high current pulse operation. The mesa
area is about 300 μm square.

Due to the high distribution coefficient of Al in the Ga solution in favour of the solid, the liquid rapidly loses Al
and consequently the Al concentration in the solid decreases
fast. Accordingly, rather steep Al profiles result as shown
in Fig. 4-9. Aluminum profiles, as displayed in Fig. 4-9, can
be measured by x-ray fluorescence analysis in a scanning
electron microscope.

The thickness of the whole epitaxial layer is about 150 μm,
of which the n type part comprises typically 30 μm. Remember
now that according to Fig. 2-4 the band gap energy of the
mixed crystal increases in proportion to the Al content.

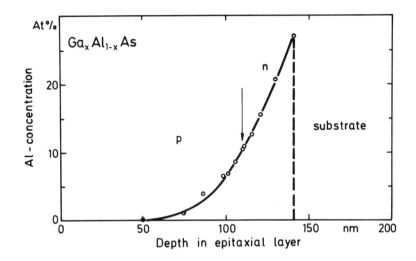

Fig. 4-9 Typical Al-concentration profile in a layer
         grown by liquid phase epitaxy for the pro-
         duction of GaAlAs:Si diodes

Therefore radiation produced at the junction of the profile
in Fig. 4-9 cannot escape to the left, where pure GaAs covers
the GaAlAs, but it cannot escape to the right either, because
there the substrate absorbs emitted radiation. The solution
to this problem is to completely eliminate the substrate and
only to keep the epitaxial layer.

The corresponding wafer process is summarized in Fig. 4-10.
The epitaxial layer has a thickness of 140 to 200 μm, with
the n side 30 to 50 μm. First, the whole p side is covered by
unstructured Au:Zn; after that the substrate is completely
dissolved in an etch consisting of a mixture of $H_2O_2$ and
$NH_4OH$ in the ratio 10:1, for example. This etch has the advantage that it selectively etches GaAs, but essentially

stops at GaAlAs with an Al concentration above about 20 %.
After metallization of the n side by Au:Ge, pattern definition, and alloying, the wafer is ready for sawing. To improve
the bondability of the Au:Ge contact it may be covered by a
pure Au layer.

Characteristics and fabrication of GaAlAs:Si diodes were
first described by DAWSON (1977). The wavelength region which
can be covered by this type of diode is indicated in Fig.
4-11. The spectrum peaking at 940 nm corresponds to a diode
of pure GaAs. By increasing the Al content the spectrum can
be shifted as shown to 828 nm. However, the half-width of the
spectrum increases rapid-
ly because the gradient
of the Al concentration
becomes steeper. Moreover
the efficiency decreases
for lower emission wave-
lengths. Therefore a
diode with a peak wave-
length between 870 and
890 nm has become fairly
standard.

Typical data of the two
diodes described are col-
lected in Table 4-1. The
efficiency of the packag-
ed GaAlAs diode at 100 mA
forward current is about
40 % higher than that of
the GaAs diode. Under

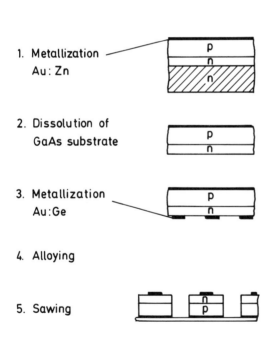

Fig. 4-10 Wafer technology for the production of GaAlAs:Si
          diodes

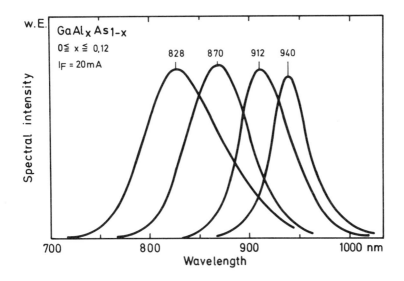

Fig. 4-11 Emission spectra of GaAlAs:Si diodes with
          varying Al content. The peak at 940 nm cor-
          responds to pure GaAs:Si, that at 828 nm to
          an Al content of about 11 % at the pn
          junction

pulsed operation at a forward current of 1.5 A the advantage
increases to more than 100 % due to the much better linearity
of the GaAlAs diode. However, it has to be mentioned that the
maximum current of the GaAlAs diode is limited to a lower
value than that of the GaAs diode. This limitation is due to
the experience that the GaAlAs diode has a tendency to degrade much more readily than the GaAs diode above a certain
current level. Unfavourable with respect to high current applications is also the much lower thermal conductivity of
GaAlAs compared with GaAs. Therefore it is wise to use GaAlAs
diodes to save current instead of pushing the output power
much beyond that reachable by GaAs diodes. An additional ad-

Table 4-1

Typical data of GaAs:Si and GaAlAs:Si diodes packaged in 5 mm
diameter clear plastic cases

| Parameter | GaAs:Si | GaAlAs:Si | unit |
|---|---|---|---|
| Maximum mean forward current | 150 | 100 | mA |
| Output power at $I_F$ = 100 mA | 12 - 18 | 16 - 26 | mW |
| Output power at $I_F$ = 1.5 A, pulsed | 110 - 160 | 210 - 350 | mW |
| Peak wavelength | 940 | 880 | nm |
| Forward voltage at $I_F$ = 100 mA | 1.25 | 1.40 | V |
| Forward voltage at $I_F$ = 1.5 A, pulsed | 1.9 | 3.0 | V |

vantage of the GaAlAs diode is its emission in a lower wavelength region, where Si photo transistors and Si based integrated detectors have a significantly higher sensitivity.
Finally a slightly higher forward voltage is usually only a
marginal disadvantage. The main reason for the higher voltage
is not the higher band gap energy but higher ohmic resistances of the bulk GaAlAs as well as of the n side contact on
the high Al concentration surface.

Several aspects of the diodes described have been treated
e.g. by BYER (1970) and later by LADANY (1971), mostly on an
empirical basis. In contrast, not much work has been published to derive a theoretical model combining electrical and
optical features and in this way allowing the optimization of

a diode for special purposes by varying design parameters on
a purely theoretical basis. Device modeling in this sense is
very incomplete. First efforts towards a solution have been
made only recently by KONTKIEWICZ (1982).

## 4-2-2 High power types

To push the power output of IREDs to higher levels one has to
consider the reasons for the power limitation in standard
IREDs or possibly existing unused reserves of power output.

The continuous wave output of standard diodes is limited by
electrical heating of the pn junction because the efficiency
decreases with increasing temperature at a rate of about
- 1 %/° C. In pulsed operation the saturation of the radiative recombination process probably becomes important besides
the heating effect. The reason for the saturation is the relatively low concentration of majority carriers due to the
high degree of compensation characteristic for Si doping,
moreover the long radiative recombination times in the band
tail states favour saturation.

A major restriction to increasing the power output without
increasing the current input exists in the form of the wide
disparity between low external and high internal efficiency
of all IREDs. This is effected by the large step of the refractive index at the diode surface, causing a very small
angle of total reflection and consequently limiting the power
externally emitted to a small fraction - about 2.7 % (air) or
7.5 % (epoxy) - of the internally produced radiation.

The limitations due to heating and saturation can be pushed
higher in a very straightforward way. By increasing the area
of the pn junction the current density can be decreased and
at the same time the thermal resistance of the diode chip

decreases. The shape of the contact has to be adjusted to
enable the current to spread evenly over the active area even
at high forward current pulses. Naturally this enlarged chip
has to be mounted on a low thermal resistance heat sink. Diodes designed according to these principles can have the following data, for example: junction area 1.2 mm$^2$; mounting polarity may be p side up as with standard diodes, but mounting
p side down results in a lower thermal resistance; thermal
resistance of the heat sink about 2 K/W. At a forward current
of 1 A one obtains 70 to 85 mW optical output power with
epoxy capsulation but without a metal reflector around the
diode chip. With such a reflector 100 to 130 mW should be
reached. These diodes can be driven up to 1.5 A continuous
wave without degrading unduly.

The second possibility, the improvement of the coupling efficiency between the interior and the exterior of the diode is
much more difficult to realize. Fig. 4-12 visualizes the escape of radiation from a diode chip. The escape is governed
by the law of refraction:

$$\frac{\sin \alpha_1}{\sin \alpha_2} = \frac{n_2}{n_1} \qquad (4\text{-}1)$$

The angle $\alpha_T$ of total reflection is obtained for $\alpha_2 = 90°$, or
$\sin \alpha_2 = 1$:

$$\sin \alpha_T = \frac{n_2}{n_1} \qquad (4\text{-}2)$$

For an epoxy capsulated diode ($n_2 = 1.5$ and $n_1 = n_{GaAs} = 3.6$)
the angle of total reflection is $\alpha_T = 25°$ and for an air capsulated diode $\alpha_T = 16°$.

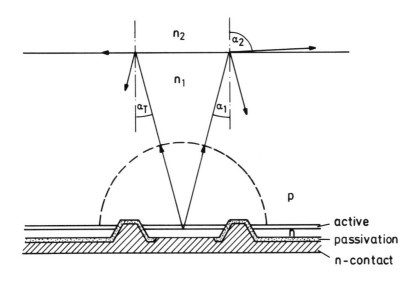

Fig. 4-12 Total reflection prevents the escape of
much of the radiation produced inside a
diode chip. A sufficiently large spherical
surface, however, avoids total reflection.

This means that only 7.6 % and 2.7 % of the internally produced radiation is transmitted through one surface to the
outside, as shown in more detail in section 6-1-3 and in the
solution to question 6-4. As indicated in Fig. 4-12 total reflection can be avoided by shaping the diode surface into a
sphere. The radius of the sphere has to be much larger than
that of the pn junction to avoid total reflection also of the
radiation emitted by the rim of the junction. The production
of this type of diode has been started by only a few manufacturers, however, because the mechanical polishing of spheres
is a very difficult and expensive process.

Sphere shaped diodes with GaAlAs double heterostructures and
hemispheres of high Al content GaAlAs have reached external
quantum efficiencies of 45 % at 50 mA forward current and

about 110 mW optical output power at 200 mA forward current
as reported by KURATA et al (1981). The maximum current of
these diodes is limited, however, due to their small emitting
areas, to 200 to 600 mA.

## 4-2-3 Optical communication types

Glass fibers are used as signal transmission lines in optical
communications. The glass fibers are either single mode or
multimode transmitters. Lasers couple power efficiently into
single mode and multimode fibers, whereas IREDs couple efficiently only into multimode fibers. The refractive index of
multimode fibers follows either a step index or a gradient
index profile. In the first case typical diameters are between 125 μm and 250 μm, whereas gradient fibers have diameters between 50 and 60 μm. These small entrance faces of fibers and the usually small aperture angles enable only high
radiance emitters to inject optical power efficiently into
fibers. IREDs for optical communication are therefore high
radiance types with large modulation bandwidths to utilize
the vast transmission capability of glass fibers.

Two classes of emitters have been developed: surface emitters
and edge emitters. Both use double heterostructures to improve efficiency by charge carrier confinement. Figs. 4-13
and 4-14 show typical examples of surface and edge emitters.
The surface emitter or BURRUS diode has a small area contact,
usually on the p side of the junction. For an emission wavelength between 800 and 860 nm the double heterostructure is
formed by an active GaAlAs layer of low Al concentration
sandwiched between two higher Al concentration layers of opposite polarity. The resistance of the small area contact,
usually made of Au:Zn, is minimized by a GaAs contact layer.

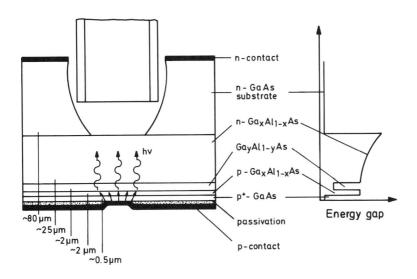

n-contact

n- GaAs
substrate

n- $Ga_xAl_{1-x}As$

$Ga_yAl_{1-y}As$

p - $Ga_xAl_{1-x}As$

p⁺- GaAs

passivation

p-contact

hv

~80 µm
~25 µm
~2 µm
~2 µm
~0.5 µm

Energy gap

Fig. 4-13 Surface emitter of the Burrus type. Three
of the four epitaxial layers form a double
heterostructure. A well has to be etched
into the GaAs substrate to remove all absorbing material above the emitting area.

All four layers are epitaxially grown from Ga solutions. Because of the absorbing GaAs-substrate a well has to be etched
into the substrate to open a path of low optical absorption
between the glass fiber and the transparent GaAlAs window.
The well, as well as the n side contact, have to be exactly
adjusted above the emitting area, defined by the p contact.
It is obvious that this type of diode requires sophisticated
processing steps.

Edge emitters are derived from laser structures and are in
some respect simpler than their surface emitting counter-
parts. In edge emitters radiation is utilized which is emit-

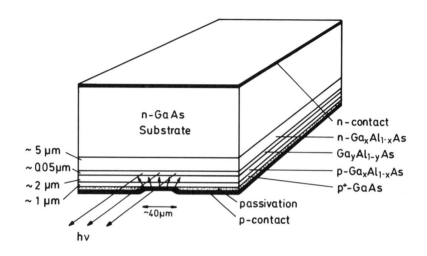

Fig. 4-14 Edge emitter similar to a laser structure.
The axis of the glass fiber is oriented parallel to the pn junction of the double
heterostructure.

ted within the junction plane. The epitaxial layer structure
is very much the same as in surface emitters. There is no
need to form a well or to have a structurized substrate contact. However there should be at least one emitting facet of
high optical quality for optimized coupling of radiation into
fibers. This is a production step usually made by careful
scribing and breaking. The other mirror facet can be etched.
Its reflectivity can be used to add to the power output. The
diode is prevented from lasing by a relatively broad contacted stripe or by unpumped absorbing material on the reverse
side.

The mutual merits of surface and edge emitters have been considered by BOTEZ and ETTENBERG (1979) and are best understood

by looking into some basic aspects of recombination theory. Modulation bandwidth and efficiency are the two most important features of IREDs for optical communication. Both features can be expressed in terms of charge carrier lifetimes. The internal efficiency $\eta_i$ as well as the total carrier lifetime $\tau$ can be expressed by radiative and nonradiative lifetimes $\tau_r$ and $\tau_{nr}$, as derived in section 6-1-2:

$$\eta_i = \frac{\tau_{nr}}{\tau_r + \tau_{nr}} \qquad (4-3)$$

$$1/\tau = 1/\tau_r + 1/\tau_{nr} \qquad (4-4)$$

The modulation bandwidth f is inversely proportional to the carrier lifetime, as shown for homojunctions in section 6-3.

$$f = \frac{1}{2\pi\tau} \qquad (4-5)$$

The lifetimes $\tau_r$ and $\tau_{nr}$ can be expressed by diode data. The inverse radiative lifetime is proportional to the total carrier density, again as shown in section 6-1-2:

$$\frac{1}{\tau_r} = B_r (n_0 + p_0 + \Delta n) \qquad (4-6)$$

where $n_0$ and $p_0$ are the equilibrium densities and the recombination coefficient $B_r$ can be derived through detailed balance arguments from the absorption coefficient of the semiconductor as explained in chapter 6. $B_r$ depends on the electronic band structure, whereas $n_0$ and $p_0$ are adjusted by doping.

The nonradiative component is dominated by the interfacial recombination velocity s at the heterostructure barriers if highly doped active regions with thicknesses d much smaller

than a carrier diffusion length are considered. Therefore
thin active layers mean short nonradiative lifetimes:

$$\frac{1}{\tau_{nr}} = \frac{1}{\tau_{nr\ bulk}} + \frac{2s}{d} \approx \frac{2s}{d} \qquad (4-7)$$

Optimizing a double heterostructure diode for high internal
efficiency means, according to Eq. (4-5), making $\tau_{nr}$ as long
as possible compared with $\tau_r$. A large nonradiative recombination time is reached by a large active layer thickness d (see
(4-7)). However, for d larger than about a diffusion length
the positive effect of the carrier confinement vanishes and
$\tau_r$ increases because $\Delta n$ decreases. Therefore the optimum
thickness d will be somewhat smaller than a diffusion length.
Consequently power optimized surface emitters have active
layer thicknesses of about 2.0 to 2.5 μm. They emit up to
15 mW optical power at a forward current of 300 mA. The angle
characteristic is Lambertian and a correspondingly small
fraction of the total output can be coupled into a fiber.

Edge emitters suffer from heavy internal absorption due to
the long path that any radiation has to cover attempting to
escape through the cleaved facet. This means that in edge
emitters the external efficiency is a smaller fraction of the
internal efficiency than in surface emitters. However, the
absorption can be drastically reduced by forcing the light
wave into the nonabsorbing cladding layers. This can be
achieved by reducing the active layer thickness, a measure
that has also the beneficial effect of giving the emitted
radiation a forward characteristic due to the fact that a
larger fraction of the emitted power travels a long distance
parallel to the junction. The forward characteristic improves
the coupling efficiency of edge emitters. Thus a power optimized edge emitter has a much smaller active layer thickness

than a surface emitter (only about 0.05 to 0.1 µm) and at the same time it has lower total output power. But due to its forward characteristic and its small emitting area the power coupled into low aperture fibers with small entrance faces can be higher than with surface emitters. Moreover, the thin active layer reduces the nonradiative lifetime and therefore increases the modulation bandwidth from 10 to 20 MHz in power optimized surface emitters to about 100 MHz in comparable edge emitters. Therefore surface emitters tend to be preferred for low data rates (f < 20 MHz) through thick fibers (d > 100 µm) with high apertures (NA > 0.25), whereas the opposite is true for edge emitters: data rates up to and beyond 100 MHz and gradient fibers with small diameters (d ≈ 60 µm) and low apertures (NA ≈ 0.18). Edge emitters, unlike surface emitters, are even considered as sources for monomode fibers for transmission over short distances.

## Questions

Q4-1   Why do some LED types have a full area, others a
       structured reverse side contact?

Q4-2   Imagine you have standard IREDs from GaAs or GaAlAs
       (see Fig. 4-7). How can you distinguish between
       both types without microscopic observation
       or accurate measurements of properties?

Q4-3   Why has a power optimized double heterostructure edge
       emitter a lower output power and a higher modulation
       bandwidth than a surface emitter?

# References


BOTEZ, D. and M. ETTENBERG, "Comparison of surface- and
    edge-emitting LEDs for use in fiber-optical communica-
    tions," IEEE Trans., ED-26 (1979), 1230

BURRUS,C.A. and B.I. MILLER, "Small area, double heterostruc-
    ture AlGaAs electroluminescent diode sources for optical
    fiber transmission lines," Optics Commun., 4 (1971), 307

BYER, N.E., "Electroluminescence in amphoteric Si-doped GaAs
    diodes I. Steady-state response," J. Appl. Phys., 41
    (1970), 1597

BYER, N.E., "Electroluminescence in amphoteric Si-doped GaAs
    diodes II. Transient response," J. Appl. Phys 41 (1970),
    1602

CRAFORD, M.G. and D.L. KEUNE, "LED technology," Solid State
    Technology, 17 (1974), 39-46, 58

DAWSON, L.R., "High efficiency graded-band-gap GaAlAs LEDs,"
    J. Appl. Phys., 48 (1977), 2485

HOFFMANN L., G. ZIEGLER, D. THEIS and C. WEYRICH, "Silicon
    carbide blue light emitting diodes with improved exter-
    nal quantum efficiency," J. Appl. Phys., 53 (1982),
    6962-6967

ISHIGURO, H., K. SAWA, S. NAGAO, H. YAMANAKA and S. KOIKE,
    "High efficient GaAlAs light-emitting diodes of 660 nm
    with a double heterostructure on a GaAlAs substrate,"
    Appl. Phys. Lett., 43 (1983), 1034-1036

IWAMOTO, M., M. TASHIRO, T. BEPPU and A. KASAMI, "High effi-
    ciency GaP green LED's with double n-LPE layers," Japa-
    nese J. Appl. Phys., 19 (1980), 2157-2163

JACOB, G., M. BOULOU, M. FURTADO and D. BOIS, "Gallium ni-
    tride emitting devices, preparation and properties,"
    Journal of Electronic Materials, 7 (1978), 499-514

KONTKIEWICZ, A.M., "The influence of design factors on the
    radiation power of GaAs:Si LEDs," Electron Technol., 13
    (1982), 81

KURATA, K., "An experimental study on improvement of perfor-
    mance for hemispherical IREDs with GaAlAs grown junc-
    tions," IEEE Trans., ED-28 (1981), 374

LADANY, I., "Electroluminescence characteristics and effici-
    ency of GaAs:Si diodes," J. Appl. Phys., 42 (1971), 654

NIINA, T., "GaP red light emitting diodes produced by a rota-
    ting boat system of liquid phase epitaxial growth,"
    J. Electrochem. Soc., 124 (1977), 1285-1289

NISHIZAWA, J.-I., Y. OKUNO, M. KOIKE and F. SAKURAI, "Bright
    pure green emission from N-free GaP LED's," Jap. J.
    Appl. Phys., 19 (1980), Supplement 19-1, 377-382

SAUL, R.H., T.P. LEE and C.A. BURRUS, "Light-Emitting-Diode
    Device Design," in Semiconductors and Semimetals, Vol.
    22, Academic Press (1985), Orlando, Fla.

STRINGFELLOW, G.B. and D. KERPS, "Green-emitting diodes in
    vapor phase epitaxial GaP," Solid State Electronics, 18
    (1975), 1019-1028

VARON, J., M. MAHIEU, P. VANDENBERG, M.-C. BOISSY and
    J. LEBAILLY, "High brightness GaAlAs heterojunction red
    LEDs," IEEE Transactions on Electron Devices, ED-$\underline{28}$
    (1981), 416-420

# 5 OPTICAL CHARACTERIZATION

## 5-1 Radiometric and photometric units

Light emitting diodes are characterized by the conversion of
electrical power into electromagnetic radiation. Therefore,
suitable units for measuring the emitted radiation are necessary and will be explained in this chapter. Radiometric units
are based only on physics, whereas photometric units create a
relation between physics and physiology, specifically the
properties of the human eye.

To start with radiometric units, we consider a source of
electromagnetic radiation. The "radiant flux" of this source
is the total energy emitted into all directions per unit
time. The SI unit for the radiant flux P is the watt.

If we want to know how the total radiant flux is composed of
flux components emitted into different directions, we have to
use another unit. We consider the source in this case from a
large distance compared with the extension of the source so
that it appears as a point source. The term which proves to
be appropriate is the "radiant intensity" J. Radiant intensity is defined as the radiant flux emitted per space angle $\Omega$.
The unit of the space angle is the steradian, abbreviated sr;
therefore we obtain the unit of the radiant intensity as
watts per steradian, in short W/sr. The unit 1 sr is defined
as the space angle formed by a cone whose tip is centered in

a sphere of radius 1 m and which cuts out of the sphere's surface an area of 1 m$^2$. Generally, for a sphere of radius r and an area A,

$$\Omega = A/r^2 \qquad (5-1)$$

Isotropic sources are defined as emitting the same radiant intensity in all directions. In general, however, the radiant intensity varies with the direction into which radiation is emitted. An IRED, for example, usually has its maximum radiant intensity in the direction of the symmetry axis and if not specified further, "radiant intensity" means this maximum intensity. Naturally, the integration of the radiant intensity over all directions of the emission results in the radiant flux of the source:

$$P = {}_\Omega\!\int J(\varphi)\; d\Omega = 2\,\pi \int_o^\pi J(\varphi)\sin\varphi \; d\varphi \qquad (5-2)$$

Here, a symmetric emission has been assumed and $\varphi$ is the angle between the symmetry axis and the direction of emission.

If we are interested in the characterization of a source not as a point source but as an extended one, there arises the question of how the total radiant flux is further decomposed into differential radiant fluxes emitted by small parts of the source in various directions. The appropriate unit to quantify the answer is the "radiance". The radiance R of a source is defined as the power radiated into a specific direction per surface element normal to this direction and per space angle. The corresponding situation is visualized in Fig. 5-1, where it is shown that as surface area only the projection dA' of the real surface dA on the plane normal to the direction of emission has to be considered. In practice, however, in most cases the emission is perpendicular to the

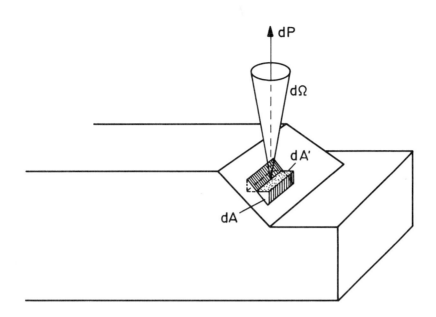

Fig. 5-1 Visualization of the definition of radiance
         R. Radiance is the output power dP per sur-
         face element dA' and per space angle dΩ,
         i.e. R = dP/dA'dΩ

emitting surface. The SI unit of the radiance is watt per
square meter per steradian, in short $W/m^2$ sr.

Considering detectors, the "irradiance" D of the detector by
a source may be encountered. It is defined as the radiant
power per surface area and measured in $W/m^2$.

All the above mentioned units may be used as wavelength derivatives. However, no special terms or units have been introduced for these derivatives, such as $\frac{dP}{d\lambda}$, measured in W/nm.

Each radiometric unit has a photometric equivalent. The keys
to convert radiometric units into photometric ones are the

"relative luminosity" function V and the definition of a photometric unit, the candle, corresponding to the radiant intensity.

The relative luminosity function V has been defined by the
Commission Internationale de L'Eclairage (CIE). It is the
average human eye sensitivity for a 2° viewing angle and daylight vision (photopic vision). For low illumination levels a
corresponding function V' has been defined (scotopic vision)
which does not, however, apply to LED characterization. Both
functions are displayed
in Fig. 5-2. For calcula-
tions tabulated values of
these functions should be
used, as published e.g.
in the DIN norm 5031.

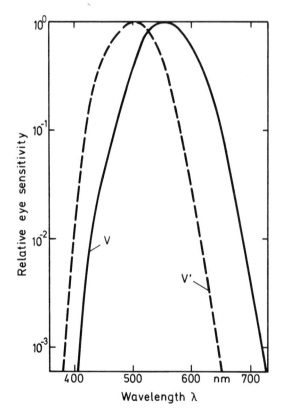

To convert a radiometric
unit X into its photome-
tric equivalent $X_V$ we
have to proceed in the
following way:

$$X_V = K \int \frac{dX}{d\lambda} V(\lambda) d\lambda \qquad (5\text{-}3)$$

In this formula K is the
photometric radiation
equivalent. It will be
shown at once how its
value is determined.

Fig. 5-2 Relative luminosity functions $V(\lambda)$ for photopic vi-
         sion (light adapted eye) and $V'(\lambda)$ for scotopic vi-
         sion (dark adapted eye). Note the logarithmic sensi-
         tivity scale

First, we introduce the photometric term "luminous flux", F, which corresponds to the radiometric radiant flux. The unit of the luminous flux is lumen (lm).

Similarly, the "luminous intensity", I, is the equivalent of the radiant intensity. Therefore the unit of the luminous intensity is lm/sr or candle (cd). The photometric radiation equivalent K is now obtained by introducing a standard for the unit cd. The luminous intensity of the standard, a black body of 1/600 000 $m^2$ area at the freezing point of platinum (2042 K) is defined as 1 candle. Applying now formula (5-3) to the standard, we find that K can be determined. The left hand side reads by definition $X_v$ = I = 1 cd and on the right hand side $\frac{dX}{d\lambda} = \frac{dI}{d\lambda}$ may be derived theoretically from the radiation laws of the black body or may be found experimentally. Because V($\lambda$) is known, the only unknown is K. Precision measurements have shown that

$$K = 673 \ lm/W \qquad\qquad (5-4)$$

for photopic vision, and

$$K' = 1725 \ lm'/W \qquad\qquad (5-5)$$

for scotopic vision. As mentioned above, for LED characterization as well as for many other purposes only formula (5-4) for daylight vision is used. By introducing the numeric value of K into formula (5-3) we have reached our goal to convert radiometric into photometric units.

The remaining photometric terms are "luminance" or "brightness" measured in the unit lm/sr $m^2$ and "illuminance" measured in lm/$m^2$, or lux (lx).

Radiometric and photometric terms and units are summed up in Table 5-1.

Table 5-1

Radiometric and photometric terms and units

| Radiometric term | Symbol | Unit | Photometric term | Symbol | Unit |
|---|---|---|---|---|---|
| Radiant flux | P | W | luminous flux | F | lm = cd sr |
| Radiant intensity | J | W/sr | luminous intensity | I | cd |
| Radiance | R | W/sr m$^2$ | luminance or brightness | B | lm/sr m$^2$ = cd/m2 |
| Irradiance | D | W/m$^2$ | illuminance | L | lm/m$^2$ = cd sr/m2 |

## 5-2 Color characterization

It might appear sufficient to characterize the color of the
light emitted by an LED by the peak wavelength of its spectrum. This, however, is not true. In a green emitting GaP:N
LED for example the main emission peak broadens on the long
wavelength side with increasing nitrogen concentration without a corresponding shift of the peak wavelength. Or, to take
another example, there may be a green emitting diode with a
red emission peak of varying strength. Thus, peak wavelength
is not a suitable measure for color. It could be used for
this purpose only in the most simple case: a spectrum consisting of a single narrow line. The appropriate concept of

color characterization has been defined by the CIE in 1931.
Within this concept any color can be described by two variables x and y. The starting points are the three color match-
ing functions $\bar{x}(\lambda)$, $\bar{y}(\lambda)$ and $\bar{z}(\lambda)$. These functions are displayed in Fig. 5-3 for a 2° viewing angle; equivalent func-
tions are defined for a larger viewing angle of 10°. For
practical purposes numerical values have to be used which can
be found in the literature, e.g. the DIN norm 5033 or the
Handbook of Colorimetry. It should be mentioned that the
color matching function $\bar{y}(\lambda)$ for 2° is identical with the
relative luminosity function V for photopic vision. The color
of an emitter with the
spectral radiant flux $\frac{dP}{d\lambda}$
can now be defined in the
following way: we start
with the three variables
X, Y and Z which are de-
fined by Eqs. 5-6 to 5-8.

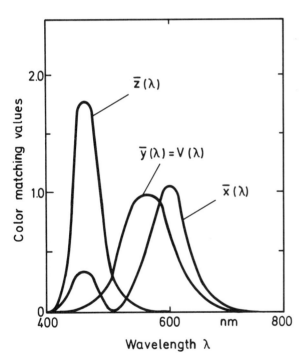

$$X = \int \frac{dP}{d\lambda} \bar{x}(\lambda)d\lambda \qquad (5\text{-}6)$$

$$Y = \int \frac{dP}{d\lambda} \bar{y}(\lambda)d\lambda \qquad (5\text{-}7)$$

$$Z = \int \frac{dP}{d\lambda} \bar{z}(\lambda)d\lambda \qquad (5\text{-}8)$$

Y is proportional to the
luminous flux as can be
seen by comparing the
expression for Y with
equation (5-3). Since the

Fig. 5-3 Color matching functions $\bar{x}(\lambda)$, $\bar{y}$ $(\lambda)$ and $\bar{z}(\lambda)$ for a
2° viewing angle; they are the basis for color de-
finition

assignment of a color to an emitter should not depend on its
luminous flux, it is reasonable to normalize the above ex-
pressions in the following way:

$$x = \frac{X}{X+Y+Z}$$ (5-9)

$$y = \frac{Y}{X+Y+Z}$$ (5-10)

The two independent variables x and y are sufficient to characterize the color of an
emitter, whereas the de-
pendent variable z = 1 -
(x + y) is no longer ne-
cessary. The possible
values of x and y form
the CIE chromaticity dia-
gram. Let us explore a
bit this diagram which is
displayed in Fig. 5-4. If
we insert into equations
5-6 to 5-8 for $\frac{dP}{d\lambda}$ a
single narrow spectral
line with wavelength $\lambda_o$,
we obtain (x, y) pairs on
the horseshoe curve, the
"spectrum locus". The
spectrum locus reaches
from about $\lambda_o$ = 400 nm to
about $\lambda_o$ = 700 nm. The

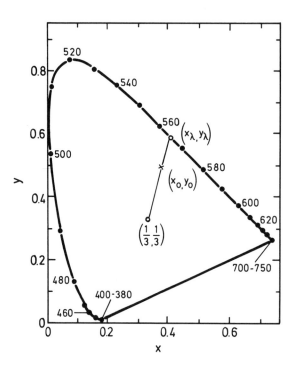

Fig. 5-4 CIE chromaticity diagram. All possible color coordinates (x, y) are on or inside the horseshoe curve.
Pure colors lie on the curve, whereas the white
point has the coordinates (1/3, 1/3)

colors on the purple line, the straight line connecting the
400 nm point with the 700 nm point, cannot be realized by a
real single spectral line, they correspond to mixtures of red
(700 nm) and blue (400 nm).

White light is characterized by $\frac{dP}{d\lambda}$ = const. If we calculate
the color coordinates for this case, we obtain $x = y = \frac{1}{3}$, the
coordinates of the white point.

Due to the construction of the color matching functions $\bar{x}$, $\bar{y}$
and $\bar{z}$ the chromaticity diagram has still other reasonable
properties. Any color $(x_o, y_o)$ has "hue" and "excitation purity". The hue is constant on the straight line connecting
$(x_o, y_o)$ with the point W = (1/3, 1/3). Thus it is obvious to
choose the intersection $(x_\lambda, y_\lambda)$ of this line with the spectrum locus for the description of hue. The point of intersection corresponds to a wavelength which is called "dominant"
wavelength and which characterizes that particular hue. Dominant wavelength is the usual measure to characterize the color of the emission of LEDs. As opposed to the (x, y) coordinates, dominant wavelength has the advantage of giving directly a notion of the color to be expected. The deviation
between peak wavelength and dominant wavelength of LEDs reaches from typically -8 nm in the green to about +8 nm in the
orange-red, as shown in Table 6-2.

Excitation purity $\varepsilon$ varies from 0 at the white point to 1 at
the point $(x_\lambda, y_\lambda)$ on the spectrum locus. Quantitatively it
is given by

$$\varepsilon = \left| \frac{(x_o, y_o) - (\frac{1}{3}, \frac{1}{3})}{(x_\lambda, y_\lambda) - (\frac{1}{3}, \frac{1}{3})} \right| \qquad (5\text{-}11)$$

All LEDs emitting in the red to yellow-green spectral range
are highly saturated. Only blue emitting types are less satu-

rated.

Beside its merits the x-y chromaticity diagram also has some
shortcomings. First, the hue on any straight line through the
white point is not exactly constant. And second, and perhaps
more important, a constant distance between two pairs of color coordinates does not correspond to the same perceived
color difference everywhere in the chromaticity diagram. For
example two just noticeably different colors correspond in
the diagram to points whose distance varies by more than a
factor of 10 depending on the location in the diagram. The
CIE tried to overcome this deficiency by introducing a trans-
formation which should retain straight lines in the x-y diagram but which would transform distances anywhere in the x-y
diagram into new distances proportional to the perceived
color difference. The result was the CIE 1960 u-v diagram
with improved color difference-to-distance proportionality,
which is, however, still far from being exact.


## 5-3 Measuring techniques

There follows a description of methods for measuring the
radiometric, photometric and color defining units which have
been introduced earlier. We emphasize methods which are reasonably accurate and reproducible and which have in principle
proved to be suited to measure large quantities of LEDs.

To compare the efficiencies of differently packaged lamps,
the unit suited best is radiant or luminous flux at a certain
current. This unit is independent of the angular halfwidth of
the emission which is due to a specific package. A suitable
method to measure the flux of a source is displayed in Fig.
5-5. A large area detector D, calibrated in A/W, is essential. The detector may be, for example, a solar cell with a

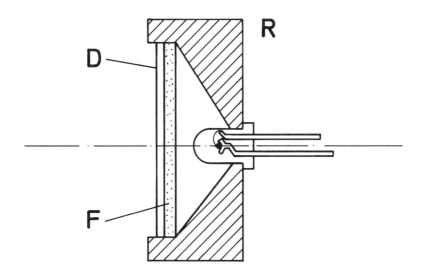

Fig. 5-5 Measuring device for radiant or luminous
         flux. Essential items are a reflector R and
         a detector D whose sensitivity curve may be
         suitably modified by the optional filter F.

diameter above about 20 mm. The optional filter F can adjust
the detector sensitivity either to a constant value for a
certain spectral range or it can produce a sensitivity profile similar to the V($\lambda$) curve. In the first case radiant flux
is measured, in the second luminous flux. The cone shaped
reflector R directs also that part of the radiation to the
detector which is emitted sideways by the diode under test.
This detector assembly has the advantages of being very com-
pact and producing a large signal. Erroneous measurements
could be due to radiation reflected by the detector and hit-
ting it again after reflection by the cone. However, this effect causes only a small correction of the calibration because a good detector will have a low reflectivity.

Radiant or luminous intensity can be measured as shown in
Fig. 5-6. A detector is used, which appears under a small
space angle d$\Omega$ when looked at from the source. According to
its definition, the intensity is obtained as the flux component falling on the detector divided by the proper space
angle. To measure the intensity as a function of the emission
angle $\varphi$ , the detector is moved around the source, or alternatively the source may be turned around the appropriate axis
without moving the detector.

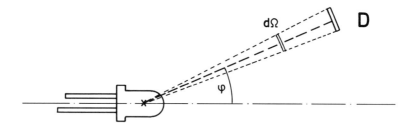

Fig. 5-6 Measurement of radiant intensity J or lumi-
           nous intensity I as a function of the emis-
           sion angle . The definitions are J = dP/d$\Omega$
           and I = dF/d$\Omega$

The third important pair of units is radiance and brightness.
These units can be measured as displayed in Fig. 5-7. A lens
L or objective is focused on the emitter's surface in such a
way that the image of a surface element of the emitter co-
incides with a detector of the small sensitive area dA. Natu-

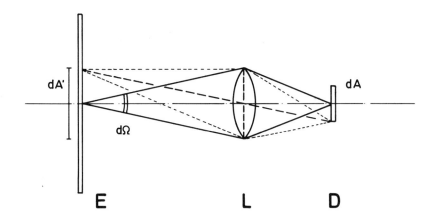

Fig. 5-7 Measurement of radiance or brightness of an
         emitter E. The aperture of the lens L de-
         fines the space angle dΩ whereas the sensi-
         tive area dA of the detector D defines the
         field of view which is identical to the
         emitting area dA'

rally the small area detector can be replaced by a diaphragm
with the aperture dA and a larger detector behind it. The
area dA is the image of an area dA' on the emitter. The aperture of the lens should be small in order to obtain a narrow
angle dΩ. The radiance R and brightness B are now obtained
according to Eqs. 5-12 and 5-13, respectively.

$$R = \frac{dP}{dA'\,d\Omega} \qquad (5-12)$$

$$B = \frac{dF}{dA'\,d\Omega} \qquad (5-13)$$

Color measurements are much more complicated than the simple
measurements described up to now. Dominant wavelength $\lambda_d$
which we use for color characterization cannot be measured
directly, because it depends on the whole emission spectrum
of an LED. Logically the simplest way to determine dominant
wavelength is to follow step by step the definition of $\lambda_d$.
This is however by no means the most economical way. Nevertheless we consider at first the tedious but precise method
and describe afterwards a less precise shortcut to arrive at
$\lambda_d$.

In the first method we start by measuring the emission spec-
trum $\frac{dP}{d\lambda}$. Here it is sufficient to use the spectrum normalized
to the highest emission peak. Then the spectrum is multiplied
by the color matching functions $\bar{x}(\lambda)$, $\bar{y}(\lambda)$ and $\bar{z}(\lambda)$ according
to Eqs. 5-6 to 5-8. The next step is the integration of the
new functions over all wavelengths, a manipulation resulting
in the three variables X, Y and Z or in numbers proportional
to them if a normalized spectrum has been chosen. LEDs emitting in the green to red spectral range have Z-values very
close to zero, thus the approximation Z = 0 may be used in
the above calculations. This is, however, not possible with
blue emitting types.

The color coordinates (x, y) are now obtained straightfor-
wardly according to Eq. 5-9 and 5-10. Finally the point
(x, y) is drawn into the chromaticity diagram and $\lambda_d$ is obtained from the point of intersection of the spectrum locus
and the straight line through the white point $(\frac{1}{3}, \frac{1}{3})$ and
(x, y). Again, some simplification is possible for green to
red emitting lamps because in this case (x, y) nearly co-
incides with the spectrum locus but this again is not true
for blue emitting LEDs.

Obviously, the procedure described above cannot be performed
quickly without using a computer. Also, the conventional
wavelength-sequential measurement of a spectrum is rather
time-consuming. Therefore it is convenient to use an optical
multichannel system which measures a whole spectrum at the
same time. Additionally the computer controlling the multichannel system can be used to calculate dominant wavelength.
Nevertheless it is not easy to build a system which is able
to perform $\lambda_d$ measurements fast, say within one or a few seconds, and certainly such a system is very expensive.

We describe now the second method, which is simple and can be
performed very quickly. The principle of this method is displayed in Fig. 5-8. The emission spectrum $\frac{dP}{d\lambda}$ needn't be spec-
trally resolved and measured. The point is that it is modi-
fied by suitable filters and only the total power contained
in the modified spectra has to be measured. Suitable filters
are filters with absorption edges coinciding with the LED
emission but having opposite slopes of the absorption curves
as shown in Fig. 5-8. The first filter may be one approxima-
ting the eye sensivity curve $V(\lambda)$, whereas many colored plastic filters can be used as the second one. The only condition
is that in the spectral range, where the filters are to be
used, reasonably steep and monotone absorption edges are present. Two measurements of the output power filtered through
the two filters create a relation to determine the dominant
wavelength $\lambda_d$. The results of the two measurements are pro-
portional to $\int \frac{dP}{d\lambda} V(\lambda)d\lambda$ and $\int \frac{dP}{d\lambda} F(\lambda)d\lambda$. The quotient Q of
both

$$Q = \frac{\int \frac{dP}{d\lambda} V(\lambda)d\lambda}{\int \cdot \frac{dP}{d\lambda} F(\lambda)d\lambda} \qquad (5-14)$$

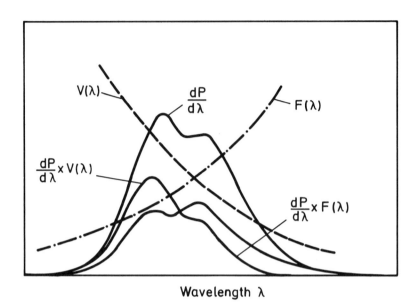

Fig. 5-8 Principle of a simple color measuring technique. The spectrum dP/dλ is modified by
filters with complementary transmission curves V(λ) and F(λ).

can be used as independent variable in a function $f_Q$ which
yields dominant wavelength $\lambda_d$:

$$\lambda_d = f_Q \ (Q) \qquad\qquad\qquad (5-15)$$

The function $f_Q$ has to be determined empirically in a calibration procedure using the color measuring technique described first. Usually $f_Q$ is very close to a linear function.

This simple method works generally only within a limited
spectral range. However, if F(λ) approximates closely $\bar{x}$(λ)
the limitation is not valid and the method is universal. The
practical advantages of the filter method have led to its
wide-spread use.

## Questions

Q5-1        What is the opening angle of a cone forming the
            space angle 1 sr, 0.1 sr and 0.01 sr?

Q5-2        What is the radiant (luminous) flux of

            a) an isotropic source, characterized by the ra-
               diant (luminous) intensity distribution
               $J(\varphi) = J_0 = $ constant?

            b) a Lambertian source, characterized by
               $J(\varphi) = J_0 \cos \varphi$ ?

Q5-3        The intensity distribution of a radiant source with
            narrow opening angle may be approximated by
            $J(\varphi) = J_0 \cos^n \varphi$ , $n \geq 2$, $n \in N$.
            Calculate the total flux.

Q5-4        Calculate some of the half intensity angles for the
            distributions of question 3, including $n = 1$.

Q5-5        Given are two narrow spectral lines of equal inten-
            sity at $\lambda_1 = 550$ nm and $\lambda_2 = 650$ nm. What are ap-
            proximately the resultant color coordinates, the
            dominant wavelength and the excitation purity?

Q5-6        What is the result of question 5 if $\lambda_2$ is changed
            to $\lambda_2 = 500$ nm?

Q5-7        What is the result of question 5 if $\lambda_1$ is changed
            to $\lambda_1 = 490$ nm?

## References

Commission Internationale de l'Eclairage (CIE)
Proc. of the 8th Sess., Cambridge (1931)

DIN (German industrial norms) 5031, Blatt 3, p.3

DIN 5033, Blatt 2, p. 2

HARDY, A.C., Handbook of Colorimetry, MIT Press (1936),
Cambridge, Mass., pp. 61-85

# 6 OPTOELECTRONIC PROPERTIES

Most important among the properties of light emitting diodes
are optoelectronic efficiency and color. The active emission
of wholly saturated colors by LEDs is always a fascinating
experience whose sensual quality reminds us of the gleaming
colors of medieval church windows. In this chapter we will
treat efficiency and color, as well as high frequency properties and reliability of LEDs.

## 6-1 Efficiency

Efficiency is defined as the ratio of output to input of an
LED measured in appropriate units. From a physical point of
view quantum efficiency is the most meaningful term. It is
defined as the number $n_p$ of photons emitted per number $n_e$ of
charge carriers flowing through the LED.

$$\eta = \frac{n_p}{n_e} \qquad (6-1)$$

'Power efficiency is the ratio of output to input power:

$$\eta_p = \frac{n_p \, h\nu \, /\Delta t}{VI} \qquad (6-2)$$

Where $h\nu$ is the photon energy, $\Delta t$ the time interval in which
$n_p$ photons are emitted, $V$ the voltage across the LED, and $I$
the forward current.

Strictly speaking $\eta_p$ characterizes not only the optoelectronic quality of a diode but depends also on purely electrical properties. For band to band transitions we have

$$h\nu = E_g = eV_g \qquad\qquad (6-3)$$

whereas the voltage V comprises the following parts

$$V = V_{pn} + I\ (R_s + R_c) \qquad\qquad (6-4)$$

Here $V_{pn}$ is the voltage across the junction and $R_s$ and $R_c$ are
the serial resistance of the semiconductor and the contact
resistance. Usually $V_{pn}$ is slightly smaller than $V_g$ and
therefore one finds for good diodes of low resistivity

$$V \approx V_g$$

In this approximation the quantum efficiency is equal to the
power efficiency.

For visual evaluation a photometric efficiency $\eta_v$ may be
introduced according to Eq. 5-3:

$$\eta_v = K \int \frac{d\eta}{d\lambda}\ V\ (\lambda)\ d\lambda \qquad\qquad (6-5)$$

The integral can be approximated by a multiplication if the
emission occurs only in a spectral range around $\lambda_o$ where $V(\lambda)$
is approximately constant.

$$\eta_v = K\ \eta\,(\lambda_o)\ V(\lambda_o) \qquad\qquad (6-6)$$

This means that the photometric efficiency is equal to the
quantum or power efficiency multiplied by the value of the
luminosity function at the wavelength $\lambda_o$.

Having defined the term efficiency, we now turn to the physics behind the concept. The external efficiency $\eta_{ext}$ of an
LED depends on several components:

(1)     The current flowing through a diode is only partly
        efficacious as a potential source of radiative recom-
        bination. This part is the diffusion current and its
        share of the total current is described as current
        injection efficiency $\eta_I$.

(2)     The minority charge carriers injected by the diffu-
        sion current recombine in part radiatively. The frac-
        tion that recombines radiatively is called internal
        efficiency $\eta_i$ and depends on the properties and the
        quality of the semiconductor material.

(3)     Not all of the photons generated will escape from the
        material; the rest will be absorbed inside the diode.
        The probability of escape is called coupling efficien-
        cy $\eta_c$.

Accordingly, the external efficiency can be decomposed:

$$\eta_{ext} = \eta_I \; \eta_i \; \eta_c \qquad\qquad (6\text{-}7)$$

Now we will consider these three components in more detail.

## 6-1-1 Injection efficiency

The properties of the current across a p n junction have been
briefly reviewed in section 1-4. As mentioned there, the diffusion current is described by

$$I = I_o \left\{ \exp\,(eV/kT)-1 \right\} \qquad\qquad (6\text{-}8)$$

Fig. 6-1 demonstrates that it is the diffusion current which
produces useful photon emission by carrier recombination
across the band gap of the semiconductor. The electrons or
holes carrying the diffusion current recombine radiatively or
nonradiatively within a few diffusion lengths around the
junction.

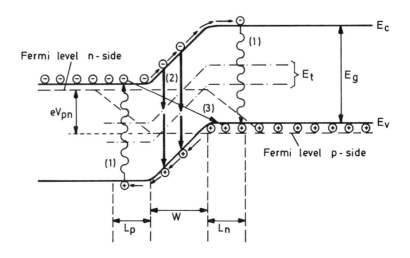

Fig. 6-1   Current components through a forward biased
           pn junction: (1) the diffusion current
           makes near band gap recombination possible,
           (2) space charge recombination current, (3)
           tunneling current. The symbols used are:
           electronic charge e, voltage across the
           junction $V_{pn}$, width of space charge region
           W, diffusion lengths $L_p$, $L_n$, energies of
           traps, band gap, valence and conduction
           bands $E_t$, $E_g$, $E_v$ and $E_c$.

However, there are other current components which are not
able to produce useful photons. These are currents caused by

charge recombination, surface recombination and tunneling.
Space charge and surface recombination both can seriously
impair injection efficiency. In Fig. 6-1 the quasi-Fermi
levels are shown under forward bias. The splitting of these
levels means that inside the space charge region the product
of electron and hole concentrations is increased beyond its
equilibrium value $n_i$ and therefore recombination must exceed
generation of carriers. All deep electron and hole traps pass
through the center energy of the two quasi-Fermi levels at
some point within the space charge region. At this energeti-
cal position the recombination rate has its maximum for each
trap. Deep traps are therefore very efficient recombination
centers inside the space charge region, whereas shallow centers are not very efficient.

The space charge current $I_{sc}$ which is due to this type of
recombination has a voltage dependency only half as strong as
the diffusion current:

$$I_{sc} \sim \left\{ \exp(eV/2kT)-1 \right\} \qquad (6-9)$$

This is because the recombination is most efficient at the
energy centered between the two quasi-Fermi levels. The recombination is mostly nonradiative, but even if it were radiative it would not produce useful photons because the traps
are not shallow.

The current component due to surface recombination has the
same voltage dependence as the space charge recombination
current. The generally observed high probability of surface
recombination is presumably caused by surface charges which
cause band bending and which in turn creates optimum recombination conditions via deep levels similarly as in the space
charge region of pn junctions. Surface recombination can only
be reduced, not prevented, by placing the junction suffi-

ciently for away from the semiconductor's surface, because
the junction somewhere meets the surface and here nonradiative recombination along the perimeter is always possible.

The current caused by tunneling carriers, also indicated in
Fig. 6-1, is highest for degenerately doped n and p sides of
the junction. It is independent of voltage within certain
limits and even decreases when the two states connected by
the tunnel process become energetically separated by the increasing forward voltage, as indicated in Fig. 6-1. The tunnel current component therefore can be completely neglected
at practical current densities, although it may be observed
at extremely low current.

To optimize injection efficiency the diffusion current has to
be made as large as possible compared with space charge and
surface recombination currents. The first possibility is to
decrease the trap densitiy by improving the material quality.
However this possibility proves to be very hard to be realized and in the case of surface recombination it is nearly
impossible. The second possibility is easily accessible
through diode design. To reduce surface recombination the
perimeter to area ratio should be small and therefore circular junction areas are favoured over, say, rectangular
areas. Additionally and most important, the difference in the
voltage dependency should be used to discriminate between
diffusion current and space charge or surface recombination
currents. Eqs. 6-8 and 6-9 tell us that the diffusion current
increases on a logarithmic current scale with twice the slope
of the other current components. Therefore a high junction
voltage has to be achieved by choosing a suitably high current density.

Experimental results obtained by PILKUHN (1981) are displayed
in Fig. 6-2. Here the electrical characteristic of a typical

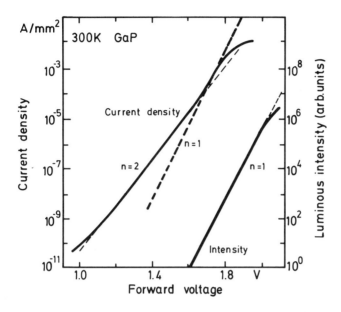

Fig. 6-2   Current density and luminous intensity of a
           GaP-diode versus forward voltage. The n=1
           and n=2 parts of the current density are
           due to the diffusion and ·space charge cur-
           rents according to Eqs. 6-8 and 6-9. The
           voltage dependency of the luminous intensi-
           ty, characterized by n=1, indicates that
           optical emission is connected with the dif-
           fusion current.

diode is plotted in a semilogarithmic current versus voltage
diagram. Fairly evident is the transition from the slope
1/2 kT (n = 2) at low voltage to 1/kT (n = 1) at higher voltages. In this example serial resistances obscure the slope
n = 1 region above about 1.8 V. Additionally the diode's
light output is plotted on the same voltage axis as the forward current. The slope 1/kT (n = 1) of this curve corresponds to the fact that only the diffusion current produces
useful radiation. Correspondingly, the more customary plots,

light output versus current, produce a superlinear dependency, as displayed in Fig. 6-3. At low current densities the potentially radiative diffusion current is only a small part of the total current. This part, however, increases with increasing forward current, producing a higher efficiency at high current levels. It is easy to show that in the limit of a negligible diffusion current the luminous intensity increases as the square of the forward current. Both the quadratic function as well as a linear function are shown as the broken lines in Fig. 6-3.

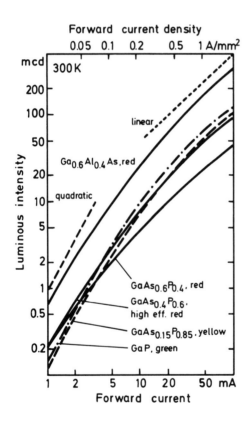

Finally the injection ratio has to be considered. It determines on which side of a pn junction radiative recombination mainly takes place. The injection ratio is defined as electron current divided by hole current. As mentioned in chapter 1, electron injection into the p side is favoured by high doping of the n side relative to the p side and by high electron mobility compared with the hole mobility. In direct gap semiconductors with high electron mobility, there-

Fig. 6-3    Luminous intensity versus forward current for several LED types. The broken lines indicate a linear and a quadratic dependency.

fore electron injection is most efficient. An example is the
GaAs:Si diode where recombination on the p side prevails. On
the other hand indirect gap semiconductors frequently have
low electron mobilities. In GaP, as an example, not only
electron mobilities are low, but usually the p side is also
highly doped and therefore the dominant light generation is
on the n side, as found by GILLESSEN and MARSHALL (1979).

## 6-1-2 Internal efficiency

The internal efficiency $\eta_i$ can be expressed as the ratio
between the radiative recombination rate $\Delta r_r$ and the total
recombination rate $\Delta r$:

$$\eta_i = \frac{\Delta r_r}{\Delta r} = \frac{\Delta r_r}{\Delta r_r + \Delta r_{nr}} \qquad (6\text{-}10)$$

where $\Delta r_{nr}$ is the nonradiative recombination rate. The $\Delta$s
indicate, according to $\Delta r = r - r_o$, the differences between
the actual rates r and their thermal equilibrium values $r_o$.
All rates are related through corresponding lifetimes to the
charge carrier density $\Delta n$ by which the density n differs due
to carrier injection from its equilibrium value $n_o$:

$$\Delta r_r = \frac{\Delta n}{\tau_r} \quad , \quad \Delta r_{nr} = \frac{\Delta n}{\tau_{nr}} \quad , \quad \Delta r = \frac{\Delta n}{\tau} \qquad (6\text{-}11)$$

As linear approximations, these relations are valid for $\Delta n$s
not too far from their equilibrium values. Using (6-10) and
(6-11) the following equations are easily derived:

$$\eta_i = \frac{\tilde{\tau}_{nr}}{\tau_r + \tau_{nr}} \qquad (6\text{-}12)$$

$$\frac{1}{\tau} = \frac{1}{\tau_r} + \frac{1}{\tau_{nr}} \qquad (6-13)$$

Equations (6-10) through (6-13) describe the fact that the internal efficiency is the result of an intense competition between radiative and nonradiative recombination processes. If nonradiative recombination is very efficient, the recombination time $\tau_{nr}$ is short and Eq. (6-12) and (6-13) are simplified to $\eta_i = \tau_{nr}/\tau_r \ll 1$ and $1/\tau = 1/\tau_{nr}$, whereas in the case of efficient radiative recombination $\eta_i = \tau_{nr}/\tau_{nr} = 1$ and $1/\tau = 1/\tau_r$ is obtained. The latter case is not very common and realized only in some infrared emitting diodes. Fig. 6-4 visualizes competing recombination processes which are either radiative and produce photons or are nonradiative and finally result in phonon production. In what follows we consider specific recombination processes, beginning with radiative recombination.

## 6-1-2-1 Radiative recombination theory

From a theoretical point of view the most satisfying treatment of radiative recombination is to start with the Hamilton operator of the crystal perturbed by an electromagnetic wave. Radiative transitions are then described by matrix elements containing the perturbation and connecting unperturbed eigenstates of the crystal. This way has been described e.g. by BEBB and WILLIAMS.
Unfortunately appropriate wave functions are not known exactly enough, nor are matrix elements with approximate wave functions easily evaluated. Therefore another method is described here, which derives a relation between absorption and emission processes. Although this method is not able to derive recombination probabilities from first principles it is nevertheless extremely useful because absorption data are readily obtained experimentally.

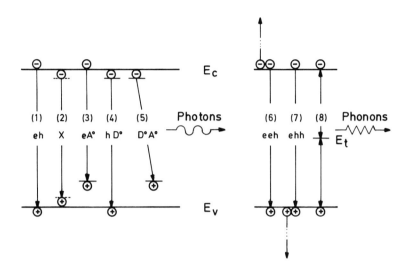

Fig. 6-4   Schematic representation of radiative and
           nonradiative recombination processes. Typi-
           cal radiative transitions are: (1) free
           electron-free hole recombination, (2) free
           exciton recombination, but the exciton can
           also be bound to an impurity, (3) free
           electron-acceptor bound hole recombination,
           (4) donor bound electron-free hole recombi-
           nation, (5) donor-acceptor pair recombina-
           tion. Typical nonradiative transitions are:
           (6) and (7) band to band Auger recombina-
           tion in n type (eeh) and p type (ehh) mate-
           rial, (8) phonon-assisted recombination at
           deep traps.

The aim is the derivation of the radiative lifetime $\tau_r$ which
is needed according to Eq. (6-12) to calculate the internal
efficiency $\eta_i$. This was first accomplished by VAN ROOSBROECK

and SHOCKLEY (1954) by considering radiative recombination
and thermal generation in thermodynamic equilibrium. In equilibrium the recombination rate $r_o$, a number per volume and
time unit, has to be balanced by the generation rate $g_o$:

$$r_o = g_o \qquad (6\text{-}14)$$

The generation rate can be calculated through Planck's law of
black body radiation:

$$P_\nu = 8\pi\, h\nu \; \frac{\nu^2}{c^2/n^2} \; \frac{1}{e^{h\nu/kT}-1} \qquad (6\text{-}15)$$

Here h is Planck's constant, $\nu$ the frequency, c/n the velocity of light inside the crystal with refractive index n, k
Boltzmann's constant and T the absolute temperature. $P_\nu$ is
the spectral irradiance or the power per frequency interval
which penetrates any unit area inside the body. The number of
photons that are absorbed per unit volume and unit time is
$\alpha(\nu)\, P_\nu /h\nu$, where $\alpha(\nu)$ is the absorption coefficient.
Therefore the generation rate is

$$g_o = \int_0^\infty \frac{\alpha(\nu)P_\nu}{h\nu}\, d\nu = 8\pi\, \frac{n^2}{c^2} \int_0^\infty \frac{\alpha(\nu)\,\nu^2 d\nu}{e^{h\nu/kT}-1} \qquad (6\text{-}16)$$

It is reasonable to assume that the non-equilibrium recombination rate $r_r$ is proportional to the product np and that $r_r$
approaches $r_o$ when equilibrium is approached:

$$r_r = r_o\, np/n_i^2 \qquad (6\text{-}17)$$

$$B_r = r_o/n_i^2 \qquad (6\text{-}18)$$

Equation (6-18) is the definition of the radiative recombination constant $B_r$; it can be calculated from Eq. (6-14) and

(6-16). Using (6-11) it is very easy now to relate $B_r$ to the radiative recombination time $\tau_r$:

$$\Delta r_r = r_r - r_o = \Delta n / \tau_r = B_r (np - n_i^2) \qquad (6-19)$$

With $n = n_o + \Delta n$, $p = p_o + \Delta p$ and $\Delta n = \Delta p$ one obtains the sought after result for $\tau_r$:

$$1/\tau_r = B_r (p_o + n_o + \Delta n) \qquad (6-20)$$

In doped material either the electron or hole concentration will predominate: $p_o \ll n_o \quad N_d$ or $n_o \ll p_o \quad N_a$, yielding for the low injection case: $\Delta n \ll N_d$ or $\Delta n \ll N_a$:

$$1/\tau_r \approx B_r N_d \quad \text{or} \quad 1/\tau_r \approx B_r N_a \qquad (6-21)$$

Therefore high doping levels make radiative recombination more efficient. However, it has to be kept in mind that above carrier concentrations of about $10^{18}$ cm$^{-3}$ the time constant of nonradiative recombination will start to decrease also.

The recombination constant $B_r$ contains the influence of the band structure of the specific semiconductor. Depending on the steepness and height of the absorption edge, $B_r$ will be large or small. In direct materials like GaAs, InP, GaSb and others it is of the order of magnitude of $10^{-10}$ cm$^3$/s. In indirect materials, where only phonon assisted absorption is possible, and absorption edges therefore are less steep and less high, $B_r$ is about four orders of magnitude smaller: $B_r$ $10^{-14}$ cm$^3$/s. Corresponding examples are GaP, and AlAs, but also Si and Ge.

6-1-2-2 Radiative recombination mechanisms

Radiative recombination channels may be classified as intrinsic and extrinsic. Intrinsic means that no centers introduced
by doping are necessary for carrier recombination, whereas
the reverse is true for extrinsic recombination. Intrinsic
recombination can be due to free carriers or free excitons;
both possibilities result in emission very close to the band
gap, as indicated in Fig. 6-4. An exciton is an excitation of
the crystal, that creates a nearly free electron and a nearly
free hole loosely bound by the electrostatic force of their
charges, very much like electron and proton in the hydrogen
atom. Due to the small effective mass of the electron and the
shielding effect of the dielectric constant of the crystal,
the binding energy is typically only about 10 meV and therefore not able to form excitons efficiently at room temperature where kT = 25 meV. In LEDs at room temperature therefore
only free carrier recombination can be important. The corresponding emission spectrum reflects in a very characteristic
way the temperature of the excited free electron system or
generally speaking of the lighter carriers. The spectral
shape is obtained as the product of the density of states in
the band and its thermal occupation probability. Therefore
the high energy side of the emission follows an exponential
$\exp(-E/kT)$ law and the half width of the emission is close to
about 2kT.

Another characteristic feature of free carrier recombination
is the shift of the emission to higher energy by a few tens
of meV at high excitation levels, as observed in pulsed operation, e.g. in GaAs IREDs. This effect is due to dynamic
band filling when the recombination rate is lower than the carrie
injection rate.

Extrinsic recombination processes make use of centers mainly introduced by intentional doping. Two important types of processes are the recombination of a free or donor bound electron with a hole bound at an acceptor. As indicated in Fig. 6-4, the emission energy of the free to bound transition is

$$h\nu = E_g - E_a + \frac{1}{2}kT \qquad (6\text{-}22)$$

Here the usually small temperature dependent term reflects the thermal energy of the free electrons similarly as in free electron-free hole recombination. The other process, bound to bound transition, is called pair recombination. The emission energy

$$h\nu = E_g - (E_d + E_a) + \frac{e^2}{\varepsilon_s r} \qquad (6\text{-}23)$$

includes a term depending on the Coulomb energy of the two centers spaced at a distance r and ionized after the recombination of the bound electron and hole. In a crystal lattice the possible values of r correspond to lattice sites and therefore are not continuous. In the case of small distances the electrostatic energy of the ionized pair reaches several tens of meV. This leads under favourable conditions to the observation of very sharp discrete emission lines. Favourable means temperatures close to absolute zero and centers that are both sufficiently deep as in GaP. Emission lines of pairs farther apart with only small Coulomb energy merge to a relatively broad emission band. In materials with shallow donors, e.g. in GaAs, only this broadened emission is observed at all temperatures. At usual LED operation temperatures, i.e. room temperature and above, all sharp emission lines disappear and pair emission bands of GaP resemble those of GaAs.

Impurities start to interact when their mutual distance approaches the diameter of the bound charge carriers. Because

the radius is large for particles with small effective mass
this happens first for donors in direct materials where the
effective electron mass is small, and consequently the donor
binding energy is also small. In GaAs for example, the
shallow donors have a binding energy of about 5 meV. Above a
doping level of about $10^{16}$ cm$^{-3}$ they are no longer able to
localize an electron, instead they form shallow tail states
below the conduction band. Shallow acceptors in GaAs have
binding energies of 30 meV, they merge at higher doping
levels around 4 x $10^{18}$ cm$^{-3}$. The appearance of tail states is
a gradual phenomenon, making the discrimination between free
to acceptor bound or tail bound states difficult. The spectra
of red emitting GaAsP-diodes and infrared emitting Zn-doped
GaAs diodes, shown in Fig. 6-5, are examples of free electron
to acceptor or tail state bound holes. These spectra are

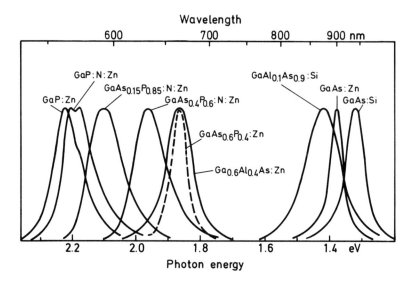

Fig. 6-5    Room temperature spectra of diodes emitting
            in the green to infrared spectral range.
            The spectra are displayed on a linear ener-
            gy scale for better comparability.

relatively narrow, the half width is about 2kT or 50 meV at
room temperature.

In materials doped with amphoteric impurities an additional
effect enhances the formation of tail states leading to deeper tails. An example of an amphoteric impurity is Si in GaAs.
Because Si is able to occupy either Ga- or As-sites, it can
act as a donor or an acceptor. Therefore both sides of the
junction in a GaAs:Si diode are usually strongly compensated
leading to a high concentration of ionized acceptors and ionized donors on both sides of the junction, leaving only a
small number of centers determining the conductivity type.
The randomly distributed charges of the ionized centers shift
the energies of the valence and conduction bands up and down
leading to a shrinkage of the effective band gap, an effect
that can be described by deep tails of the bands. The effect
is visualized in Fig. 6-6. The up and down of the band energies reduces the mobility of the charge carriers in these
tail states and because of the spatial separation between the
conduction band minima and the valence band maxima the recombination times are long. Nevertheless the radiative recombination probability is large, leading to a very efficient radiative recombination whose spectrum is broad and shifted
away from the band gap.

This spectrum is displayed in Fig. 6-5 as well and denoted by
GaAs:Si. The GaAlAs:Si spectrum is additionally broadened by
a usually large gradient in the Al concentration.

The low efficiency of indirect gap materials can be increased
by more than an order of magnitude by nitrogen doping. Nitrogen substitutes for P in GaP and $GaAs_{1-x}P_x$ and is the most
important example of an isoelectronic trap. Isoelectronic
centers can bind a charge carrier if there is a large difference in the electronegativity between the center and the
atom it substitutes for. In the case of N in GaP the larger

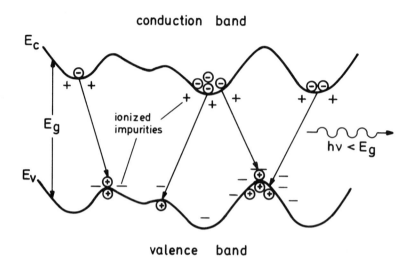

Fig. 6-6   Model of tail state formation in strongly
           compensated material. The fluctuations of
           the band energies are caused by the Coloumb
           energies of randomly distributed charged
           impurities.

electronegativity of N than that of P induces the binding of
an electron. The binding force corresponds to a short range
potential very different from the long ranging Coulomb poten-
tional of a usual donor. Consequently the electron is tightly
bound to the isoelectronic center and according to the
Heisenberg uncertainty relation its wave function spreads out
in k-space and has an appreciable amplitude at $k = 0$, although the material considered is indirect. Therefore the
recombination probability with a hole bound around $k = 0$ is
strongly enhanced. The electron can bind a hole through their
mutual Coulomb attraction. The N-bound exciton, as the entity
is called, induces according to the radiative recombination

theory, outlined in section 6-1-2-1, very strong absorption close to the intrinsic absorption edge of GaP. This has to be considered in LED design.

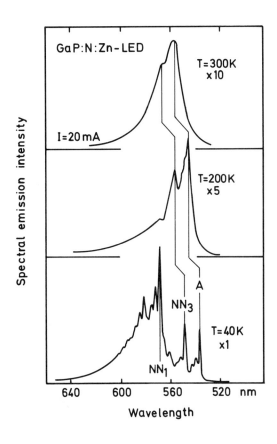

An emission spectrum caused by the N-bound exciton is displayed in Fig. 6-5. A typical feature is the double peaked structure of this spectrum. This structure is explained by the fact that excitons can be bound not only by isolated N atoms, but also by pairs of N atoms with varying separation. In Fig. 6-7 the emission of an N-doped GaP LED is shown as a function of temperature (SCHAIRER 1979). At low temperatures the double peaked structure is resolved

Fig. 6-7    Analysis of the emission spectrum of a GaP:N:Zn-LED by variation of the temperature. At low temperature zero phonon lines labeled A, $NN_3$ and $NN_1$ dominate the emission. They are accompanied on the long wavelength side by phonon replicas. The A line is caused by the recombination of excitons bound to isolated N atoms, the others are due to excitons bound to third nearest and next neighbor N pairs. With increasing temperature the lines shift to lower energy according to the band gap variation.

into a line spectrum whose main features can be explained by
excitons bound to isolated nitrogen, next neighbour pairs $NN_1$
and third nearest pairs $NN_3$. Because the abundance of the
various pairs strongly depends on the doping level, the relative intensity of the closest pair line $NN_1$ will increase
with the doping level. In this way it is possible to shift
the emission color of N-doped GaP LEDs from yellow-green to
pure yellow merely by increasing the N concentration. An example of spectra of differently doped diodes is shown in Fig.
6-8.

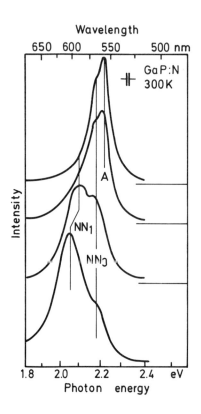

A disadvantage of N-doped
GaP diodes is the shift
of the emission away from
the band gap by the bind-
ing energy of the exci-
ton. This shift causes
the emitted color to
change from pure green to
the well known yellow-
green of most GaP LEDs.
Recently efforts have
been made to strongly
reduce nonradiative re-
combination in liquid
phase epitaxial GaP and
thus to achieve an ac-
ceptable efficiency even
without N doping. From
very high quality mate-

Fig. 6-8   Cathodoluminescence spectra of GaP with nitrogen
doping levels increasing from about 1 x $10^{18}$ $cm^{-3}$
(top) to about 1 x $10^{20}$ $cm^{-3}$ (bottom). Adapted from
results by HART (1973).

rial diodes have been prepared with pure green emission and about one third the efficiency of N-doped material. Although the wavelength difference between the two types with and without N doping is only small, as shown in Fig. 6-5, the difference in color is impressive.

6-1-2-3 Nonradiative recombination

The competition between nonradiative and radiative recombination processes leads generally to lower internal efficiencies at shorter wavelengths. Infrared emitting Si doped GaAs and GaAlAs diodes reach internal efficiencies above 50 % whereas yellow and green emitting GaAsP diodes stay below 1 %. Theoretical models have been developed for a multitude of nonradiative processes. However not yet enough is known about the question which of them are realized to what extent in any specific material.

At high doping levels the intrinsic Auger effect becomes important and possibly dominates nonradiative recombination. The Auger effect is a three particle process in which the recombination energy of an electron hole pair is transferred to a third particle, electron or hole exciting that particle deep into the respective band. There, it thermalizes rapidly by the emission of phonons (quantized lattice vibrations). Due to its three particle nature the recombination rate of the Auger process depends, by analogy to Eq. 6-17, on the third power of particle concentrations:

$$r_{nr} = C_n \, n^2 p, \text{ ehe-process} \qquad (6\text{-}24)$$

$$r_{nr} = C_p \, p^2 n, \text{ ehh-process} \qquad (6\text{-}25)$$

$C_n$, $C_p$ are Auger coefficients for n type and p type materials. Using Eq. 6-11, lifetimes can be derived

$$\tau_{An} = 1/C_n \, n^2 \qquad\qquad (6-26)$$

$$\tau_{Ap} = 1/C_p \, p^2 \qquad\qquad (6-27)$$

Auger processes can be experimentally identified by this quadratic dependence of the lifetime from particle concentrations. The decrease of efficiency generally observed above
doping levels between a few times $10^{18}$ cm$^{-3}$ and $1 \times 10^{19}$ cm$^{-3}$
is most likely caused by beginning dominance of Auger recom-
bination.

Deep level recombination is usually nonradiative, but even if
it should be radiative the photons emitted would not be useful because they are of too low energy. Nonradiative recombination is favoured, because the crystal lattice around a deep
impurity takes part in the binding of an electron or hole.
This means the equilibrium position of the lattice atoms
around the impurity strongly depends on whether a particle is
bound or not. Consequently in deep level recombination the
lattice adjusts itself to the new equilibrium position by
intense phonon emission including no photons, or only photons
of much reduced energy. Therefore, but also because deep
levels reduce the injection efficiency (see section 6-1-1),
their avoidance is extremely important although not easily
achieved.

Dislocations usually produce a region of low efficiency
around the dislocated lattice atoms. This can be seen directly in LEDs looking through a microscope on the light emitting
face of the chip or it can be detected by spatially resolved
photoluminescence or cathodoluminescence. The reason is not
well understood. It seems that a dislocation is not directly
effective in nonradiative recombination. Possibly it becomes
decorated by impurities during crystal growth and these impurities produce low quality material of low efficiency
around the dislocation. Experimentally it has been found that

it is especially important to use GaP substrates of low dislocation density to fabricate green emitting LEDs of high
efficiency by liquid phase epitaxy. In infrared emitting
GaAs:Si diodes, however, dislocation density is of much less
importance.

Surface or more generally interface recombination is another
important nonradiative recombination mechanism. The surface
recombination velocity is very high in GaAs und GaP, but sur-
prisingly low in InP, whose other properties are very similar
to GaAs. Double heterostructures incorporating interfaces of
the system GaAs/GaAlAs are efficient because the recombination velocity at these interfaces is very much reduced compared with surfaces to air. Many models have been put forward
to explain the efficient surface recombination which is generally observed. We mention only space charge recombination in
the surface layer, which is induced by band bending. In LED
design the injected charge carriers have to be kept away from
the surface by at least several diffusion lengths. However,
as mentioned in section 6-1-1, the pn junction always meets a
surface at its circumference. Therefore nonradiative surface
recombination cannot be prevented completely.

## 6-1-3 Coupling efficiency

Coupling efficiency $\eta_c$ can be very low; it is therefore an
important problem to improve it. To understand this one has
to consider total reflection and reflectivity in materials
with high refractive index. The angle of total reflection $\alpha_T$
is obtained from the law of refraction $\sin \alpha_1 / \sin \alpha_2 = n_2/n_1$
for $\alpha_2 = 90°$ (n, refractive index, the indices refer to the
two media considered):

$$\sin \alpha_T = \frac{n_2}{n_1} \qquad\qquad (6-28)$$

The refractive index of e.g. GaAs and GaP is high, $n_1$ = 3.6
and $n_1$ = 3.3 respectively. Consequently one obtains for a
semiconductor-air interface $\alpha_T$ = 16.1° for GaAs and $\alpha_T$ =
17.6° for GaP. The transmissivity T reaches its maximum for
perpendicular incidence (R, reflectivity):

$$T = 1 - R = 1 - \frac{(n_1 - n_2)^2}{(n_1 + n_2)^2} = \frac{4n_1 n_2}{(n_1 + n_2)^2} \qquad (6-29)$$

Therefore even the maximum of the transmission reaches only
0.68 for GaAs and 0.71 for GaP in the case of a semiconductor
air interface.

An upper limit for the fraction f of radiation which escapes
from a surface in a first try is easily obtained assuming
isotropic angular distribution of the radiation produced inside the crystal and transmissivity as for perpendicular incidence:

$$f \leq T \frac{\Omega_{\alpha T}}{\Omega_{90}} \qquad (6-30)$$

Here $\Omega_{\alpha T}$ is the space angle limited by total reflection and
$\Omega_{90}$ is the space angle of a hemisphere. One obtains for air
and GaAs f $\leq$ 0.68 x 0.039 = 2.7 x $10^{-2}$ and for GaP f $\leq$ 0.71 x
0.047 = 3.3 x $10^{-2}$ (see question 4, this chapter). These
values are disturbingly small. One of the important questions
in LED design is therefore: how can we improve the escape
probability?

The first answer is to avoid or reduce total reflection.
Non-planar surfaces like hemispheres offer a solution if they
are sufficiently far away from the radiation producing pn
junction. A specific type of a sphere shaped diode has been
described in section 4-2-2 and shown in Fig. 4-9. The radiation impinges nearly perpendicularly on the semiconductor

surface, therefore reflection losses are also minimal. However it is clear that this solution is very expensive and can
be chosen, in spite of its efficiency and elegance, only in
few special applications.

A very remarkable reduction, although not avoidance, of total
reflection can be reached by suitable encapsulation. For this
purpose nearly exclusively plastics of the epoxy type are
used. The refractive index of all the different varieties is
very close to 1.5. Note, that total reflection at the epoxy
surface can be circumvented in a quite natural way by the
usually lens shaped surface of the epoxy. Using Eq. (6-29)
and (6-30) we can estimate the improvement achieved by epoxy
encapsulation: $f \leq 0.83 \times 0.091 = 7.6 \times 10^{-2}$ for GaAs and $f \leq$
$0.86 \times 0.109 = 9.4 \times 10^{-2}$ for GaP. In both cases the improvement is by nearly a factor of three. Therefore epoxy encapsulation should be considered as an integral part of the LED
design and not as a more or less ornamental addition.

There is a second way to improve escape probability. That
answer is to make better use of the totally reflected radiation. What happens to this part of the internally produced
radiation? It is weakened by

(1)   volume absorption before it again reaches a surface
(2)   absorption at a surface

(3)   volume absorption after possibly multiple reflec-
      tions

To appreciate the last point one has to be aware of the fact
that a perfectly square chip will not change the angles of
reflection in such a way that total reflection could finally
be overcome. The radiation trapped inside the crystal will be
given a chance of escape only if there are irregularities at
the crystal surface which change the angles of reflection.

Therefore etching a diode chip's surface rough produces up to
a factor of 2 higher external efficiency in air and 1.6 encapsulated. In many cases, however, irregularities introduced
unintentionally during fabrication are sufficiently efficient
in randomizing the angles of reflection.

Surface absorption is mainly due to absorbing contacts. This
source of loss is minimized by choosing front contacts as
small as compatible with bonding technology and a sufficiently uniform distribution of the current at the maximum intended forward current. The reverse side contact should have a
large fraction of high reflective area. Contact types with
high reflectivities have been described in chapter 4 for
specific LEDs.

Sawn surfaces are another source of high absorption. Sawing
debris of submicron dimensions probably covers these surfaces
and causes absorption. Sawing cuts therefore have to be
cleaned up by etching. This produces at the same time useful
surface irregularities.

Finally volume absorption should be made small. Sources of
volume absorption are: band-to-band absorption, free carrier
absorption and absorption by impurities.

Band-to-band absorption is most severe in direct gap materials. Radiation produced by band-to-band recombination is
usually absorbed after only a few microns. Although absorbed
radiation can be re-emitted, an effect called photon recy-
cling, most of this radiation is lost due to usually low internal efficiency. An example which comes very close to the
conditions described are Zn-diffused GaAs diodes. Here the
emitted radiation is produced in shallow tail states introduced by the high Zn concentration. The absorption length is
below 5 μm. Therefore all of the radiation emitted in direc-

tions away from the surface will be lost. The remaining fraction can be maximized by choosing a junction depth of about 3
diffusion lengths of the electrons as has been shown e.g. by
REICHL et al (1978). This condition realizes an optimum com-
promise between nonradiative surface recombination and volume
absorption.

Much higher efficiencies can be achieved in direct gap materials when the photon energy of the emitted radiation can be
shifted away from the gap as in Si-doped GaAs due to the formation of deep tail states. In this case the whole chip becomes transparent and free carrier absorption dominates the
losses. Therefore low doping levels of the substrate further
improve the escape probability.

Band-to-band absorption is much smaller in indirect gap materials as e.g. in GaP than in direct gap materials. The absorption length is larger than typical chip dimensions and
again the chip appears transparent. The escape probability
can be improved using randomizing reflections as described
above. Quantitatively transparent chips have been considered
by JOYCE et al (1974). Other examples of these diodes with
low volume absorption are the yellow and orange emitting
GaAsP diodes on GaP substrates.

Impurities are the most important absorption source in all
diodes on GaP substrates. As described in section 6-1-2-2
nitrogen doping produces very high absorption close to the
indirect band gap. Therefore the nitrogen-doped volume has to
be made as small as possible, i.e. the doped region has to be
restricted to a few diffusion lengths on both sides of the
junction.

Finally the influence of internal absorption on angular distribution should be mentioned. Chips that are highly trans-

Table 6-1

Characteristic features of commercial LEDs. $\lambda$ and $\Delta\lambda$ are peak
wavelength and spectral halfwidth, $\eta$ the power efficiency, P
and F the radiant and luminous flux, V  the forward voltage
and $\tau$ the total carrier lifetime.

| Type | $\lambda$ (nm) | $\Delta\lambda$ (nm) | $\eta$ (%) | at 20 mA | | | $\tau$ (ns) |
| | | | | P (mW) | F (mlm) | V (V) | |
| --- | --- | --- | --- | --- | --- | --- | --- |
| GaAs:Si | 950 | 50 | 12 | 3 | – | 1.3 | 500 |
| GaAs:Zn | 900 | 20 | 4 | 1 | – | 1.3 | 50 |
| GaAlAs:Si | 870 | 80 | 18 | 5 | – | 1.4 | 500 |
| GaAlAs:Zn red | 650 | 20 | 3 | 1.1 | 80 | 1.8 | 50 |
| GaAsP:Zn red | 660 | 20 | 0.6 | 0.19 | 8 | 1.6 | 50 |
| GaAsP:N:Zn orange | 625 | 40 | 0.6 | 0.24 | 50 | 2.0 | 100 |
| GaAsP:N:Zn yellow | 590 | 40 | 0.1 | 0.04 | 20 | 2.2 | 100 |
| GaP:N:Zn green | 565 | 40 | 0.2 | 0.1 | 60 | 2.4 | 400 |
| GaP:Zn pure green | 560 | 30 | 0.07 | 0.03 | 20 | 2.4 | 500 |
| GaN blue | 490 | 80 | 0.01 | 0.01 | 1.4 | 5 | ? |
| SiC blue | 480 | 50 | 0.02 | 0.016 | 1.5 | 4 | ? |

parent produce an angular characteristic outside the chip
which is nearly isotropic, whereas chips that are absorbent
emit radiation with an angular dependency proportional to
cos $\varphi$ , i.e. they produce a Lambertian characteristic.

External efficiencies and some other characteristics of good
commercially available products are summarized in Table 6-1.

## 6-2 Emission spectrum and color

Emission spectra of diodes emitting in the visible spectral
region have been shown in Fig. 6-5. In section 5-3 it was
explained why color cannot be characterized simply by peak
wavelength. Here we present in Table 6-2 typical peak wavelengths and dominant wavelengths which describe the color of
a diode. In the green and yellow spectral region the color
sensitivity of the eye is exeedingly high. The emission spectra are strongly influenced by variations of the N concentration, by pn junction depth (through reabsorption) and crystal
composition in the case of GaAsP. Therefore green and yellow
emitting diodes can be purchased selected according to dominant wavelength. Typical step sizes are 2 or even 1 nm. The
difference between dominant and peak wavelength is minimum in
the yellow spectral range and changes sign going from there
to the green or red spectral region. The color sensitivity
strongly decreases in the orange and red spectral range and
therefore color selection is not necessary for these diode
types.

Table 6-2

Diode types of different emission colors with pertinent peak
wavelengths and dominant wavelengths.

| Diode material | Color | Peak wavelength (nm) | Dominant wavelength (nm) |
|---|---|---|---|
| GaP | green | 558 | 563 |
| GaP:N | yellowish green | 561 | 569 |
| $GaAs_{0.15}P_{0.85}$:N | yellow | 590 | 589 |
| $GaAs_{0.4}P_{0.6}$:N | red orange | 630 | 623 |
| $GaAs_{0.6}P_{0.4}$ | red | 660 | 652 |
| $GaAl_{0.4}As_{0.6}$ | red | 650 | 643 |

## 6-3 High frequency response

Most LEDs for standard application are homojunction types.
Double heterostructures are almost exclusively used for light
wave transmission through glass fibers. The high frequency
response of the sources for fiber transmission is most important, but complicated, because it strongly depends on the
current density. Below, the frequency response of homojunction diodes is treated theoretically and compared with experimental results. The theoretical treatment then forms the
basis of a discussion of the response of heterojunction
types.

The optical power P per unit area of a diode can be expressed, using (6-11), as

$$P = \int_{0}^{\infty} h\bar{\nu} \frac{\Delta n(x,t)}{\tau_r} dx \qquad (6-31)$$

Here, h is Planck's constant, $\bar{\nu}$ the average photonfrequency, $\Delta n$ the non-equilibrium charge carrier density which is time and space dependent and $\tau_r$ is the radiative recombination time. Strictly speaking P is the internally generated power density, to obtain the emitted power P has to be multiplied by $\eta_I$ and $\eta_c$, the injection and coupling efficiency. Separation into a steady state part $P_0$ and a time-varying component $P_1$ is possible by introduction of a corresponding separation of $\Delta n(x,t)$:

$$\Delta n(x,t) = \Delta n(x) + n_1(x,t) \qquad (6-32)$$

$$P_0 = \frac{h\bar{\nu}}{\tau_r} \int_{0}^{\infty} \Delta n(x) dx \qquad (6-33a)$$

$$P_1 = \frac{h\bar{\nu}}{\tau_r} \int_{0}^{\infty} \Delta n_1(x,t) dx \qquad (6-33b)$$

The charge carrier distributions $\Delta n$ and $\Delta n_1$ fall from their maximum values at $x = 0$ to low values within a diffusion length L:

$$\Delta n(x) \quad = \Delta n(0) \exp(-x/L) \qquad (6-34a)$$

$$\Delta n_1(x,t) = \Delta n(0,t) \exp(-x/L^*) \qquad (6-34b)$$

A frequency dependent, complex diffusion length $L^*$ is used to describe the fact that charge carrier injection across the pn junction becomes inefficient at high frequencies:

$$L^{*2} = \frac{L^2}{1+j\omega\tau} \qquad (6-35)$$

where $\tau$ is the total carrier lifetime. L* can be derived from the time dependent solution of the diffusion equation of the injected carriers. This equation contains exactly the same parts as Fick's second law (3-12), but includes one more term $\Delta n(x,t)/\tau$ describing carrier recombination. The integration of equations (6-33) is easy:

$$P_o = \frac{h\bar{v}}{\tau_r} \; L \; \Delta n(0) \qquad\qquad (6\text{-}36a)$$

$$P_1 = \frac{h\bar{v}}{\tau_r} \; L^* \; \Delta n_1(0,t) \qquad\qquad (6\text{-}36b)$$

It is useful to replace the charge carrier densities by the experimentally easily accessible current densities $i_o$ and $i_1$. Because the current is transported by diffusion, the law of diffusion (3-9) can be applied, yielding in connection with Eq. (6-34)

$$i_o = -eD \; \left(\frac{\partial n}{\partial x}\right)_{x=0} = eD \; \frac{1}{L} \; \Delta n(0) \qquad\qquad (6\text{-}37a)$$

$$i_1 = -eD \; \left(\frac{\partial n_1}{\partial x}\right)_{x=0} = eD \; \frac{1}{L^*} \; \Delta n_1(0,t) \qquad\qquad (6\text{-}37b)$$

The Einstein relation (1-2): $L = \sqrt{D\tau}$ and the expression for the internal efficiency $\eta_i = \tau/\tau_r$ yield:

$$P_o = h\bar{v} \; \eta_i \; i_o/e \qquad\qquad (6\text{-}38a)$$

$$P_1 = h\bar{v} \; \eta_i \; \frac{L^*}{D} \; i_1/e = \frac{h\bar{v} \; \eta_i \; i_1}{e(1+j\omega\tau)} \qquad\qquad (6\text{-}38b)$$

The first important result, the modulation efficiency $P_1/i_1$, is now directly obtained as

$$\left| \frac{P_1}{i_1} \right| = \frac{P_o}{i_o} \frac{1}{(1+\omega^2 \tau^2)^{1/2}} \qquad (6-39)$$

The modulation bandwidth $\omega_{max}$ can be defined as the frequency at which the square of the modulation efficiency has dropped to one half of its static value $P_o/i_o$. Eq. (6-39) yields for this definition simply

$$\omega_{max} = 1/\tau \qquad (6-40)$$

$$f_{max} = 1/2\pi\tau \qquad (6-41)$$

and that means the bandwidth is the inverse of the charge carrier lifetime and the bandwidth increases as the lifetime is reduced.

The static output power $P_o$ contains lifetimes according to $P_o = h\bar{\nu} \eta_i i_o/e = h\bar{\nu} \tau i_o/\tau_r e$. To analyse the implications of the above result it is useful to show explicitly the dependency of $P_i/i_1$ on the radiative and nonradiative time constants $\tau_r$ and $\tau_{nr}$. Using (6-13) one obtains

$$\left| \frac{P_1}{i_1} \right| = \frac{h\bar{\nu}}{e} \frac{\tilde{\tau}_{nr}}{\tau_r + \tau_{nr}} \frac{1}{(1+\omega^2(\frac{\tilde{\tau}_r \tilde{\tau}_{nr}}{\tau_r + \tau_{nr}})^2)^{1/2}} \qquad (6-42)$$

The asymptotic behavior of the modulation efficiency for low and high frequencies is most easily obtained from Eq. (6-42). The usually realized case of low internal efficiency, $\tilde{\tau}_{nr} \ll \tilde{\tau}_r$, yields the results

$$\left|\frac{P_1}{I_1}\right|_{\omega \to 0} \xrightarrow{\phantom{xx}} \frac{h\bar{\nu}}{e}\; \frac{\tilde{\tau}_{nr}}{\tau_r} \tag{6-43a}$$

and
$$\left|\frac{P_1}{I_1}\right|_{\omega \to \infty} \xrightarrow{\phantom{xx}} \frac{h\bar{\nu}}{e}\; \frac{1}{\omega \tau_r} \tag{6-43b}$$

High internal efficiency: $\tau_r \ll \tau_{nr}$ gives similarly

$$\left|\frac{P_1}{I_1}\right|_{\omega \to 0} \xrightarrow{\phantom{xx}} \frac{h\bar{\nu}}{e} \tag{6-44a}$$

and
$$\left|\frac{P_1}{I_1}\right|_{\omega \to \infty} \xrightarrow{\phantom{xx}} \frac{h\bar{\nu}}{e}\; \frac{1}{\omega \tau_r} \tag{6-44b}$$

Most interesting is the high frequency fall off. Independently of the total carrier lifetime $\tau$ it is given by the radiative recombination time $\tau_r$. The high frequency power output
is therefore not increased if an increase of the bandwidth
according to Eq. (6-40) is achieved through a reduction of
the nonradiative time constant. A reduction of the nonradiative lifetime only reduces the low frequency power output,
thus yielding in some way only an apparent increase of the
bandwidth. The situation is visualized in Fig. 6-9, where as
an example the frequency responses of diodes with a radiative
lifetime of 10 ns and varying nonradiative lifetimes are
shown. It is clearly seen that a reduction of the nonradiative time constant increases the bandwidth but does not mean
any more power, say, at 1 GHz.

As has been shown in section 6-1-2-1 the radiative time constant can be decreased by high doping levels according to

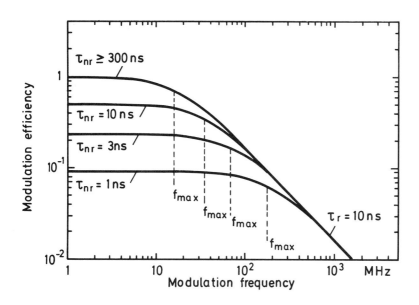

Fig. 6-9    Theoretical frequency responses of homo-
            junction diodes with a radiative lifetime
            $\tau_r$ = 10 ns and nonradiative lifetimes de-
            creasing from $\infty$ to 1 ns.

Eq. (6-20) and (6-21). High frequency diodes, therefore, have
to be doped into the $10^{18}$ $cm^{-3}$ range. Beyond a few times
$10^{18}$ $cm^{-3}$ the nonradiative time constant starts to decrease
too. A rough estimate of the high frequency fall off is obtained from Eq. (6-21) with $N_d$ = $10^{18}$ $cm^{-3}$ and $B_r \approx$
$10^{-10}$ $cm^3/s$ for direct material, e.g. GaAs. These figures
yield $\tau_r$ = 10 ns. A diode with $\tau_r$ = 10 ns and an internal
efficiency close to 1 and therefore $\tau \approx \tau_r$ would have a
bandwidth according to (6-41) of only $f_{max}$ = $1/2\pi\tau$ = 16 MHz.
Diodes of higher bandwidth are usually only obtained at the
expense of lower internal efficiency. Nevertheless this can
be useful in order to have a nearly constant modulation efficiency over the intended bandwidth range.

To judge the quality of a high frequency diode as well as the
bandwidth the low frequency efficiency should always be
known. These two figures characterize a diode's high frequency response completely, because they contain $\tau_r$ and the
ratio $\tau / \tau_r$. As a single figure of merit the bandwidth-ef-
ficiency product has been proposed.

This product corresponds to $1/ \tau_r$ and is therefore, although
better than the bandwidth alone, not sufficient to characterize the high frequency response completely.

The frequency response according to (6-39) has been frequently confirmed experimentally. As an example measurements by
MABBIT and GOODFELLOW (1975) are displayed in Fig. 6-10,
together with the fitted curve according to (6-39). In this

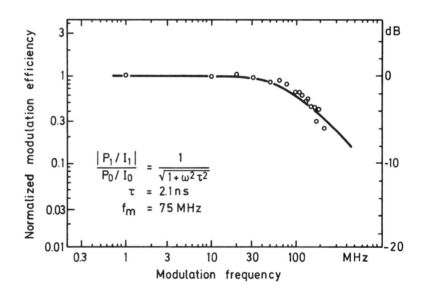

Fig. 6-10 Experimental high frequency response of a
          diode according to data of MABBIT and GOOD-
          FELLOW (1975). Theoretical fit by equation
          (6-39).

specific case the bandwidth of 75 MHz implies a total carrier
lifetime of 2.1 ns. A larger bandwidth would be obtained with
another definition of bandwidth as given in question Q6-5.

The frequency response described above applies to homojunction diodes, although quite similar responses have been found
experimentally for heterostructures. In heterojunction diodes
the assumption of unhindered carrier diffusion is not valid.
Instead, the double heterostructure confines the injected
carriers to low band gap material whose thickness is smaller
than, or even small compared with, the diffusion length. As
discussed in section 4-2-3, the nonradiative lifetime is then
influenced or even dominated by heterointerface recombination
according to Eq. (4-5). Additionally the radiative lifetime
becomes current dependent through the increasing importance
of the injected carrier densitiy $\Delta n$, see Eq. (4-4). Quantitatively one finds that the time constant is proportional to
the inverse of the square root of the current density. Therefore the frequency response of heterojunction diodes depends
much more on specific design parameters and application conditions as the one of homojunction diodes. Nevertheless some
general consequences are qualitatively summarized as follows:

(1)   As in homostructures it is most favourable to re-
      duce $\tau_r$ in order to get a large bandwidth and at
      the same time a high output power.

(2)   This can be achieved by high doping levels as in
      homostructures and with the same limitations.

(3)   High injection conditions are attractive, because
      then according to (4-4) $\tau_r$ approaches $1/B_r \Delta n$. Non-
      linear distortion of the radiative output is, how-
      ever, to be expected unless the modulation current
      is only a small fraction of the static current.

(4)  High injection can be reached at lower current if a
     small junction area and a thin active layer is de-
     signed into the diode.

(5)  As in homostructures the bandwidth increases if the
     nonradiative time constant is reduced. This can be
     achieved either by excessively high doping levels, or
     now specifically for heterostructures according to
     (4-5) by a very thin active layer.

(6)  The reduced efficiency due to a very thin active
     layer can in edge emitters in part be compensated
     for by the resulting forward characteristic of the
     optical output.

Through the measures (3) to (6), specific for double heterostructures, it is possible to reach higher bandwidths and
larger high frequency powers than with homojunction diodes.
Minus-three-dB-bandwidths up to and beyond 1 GHz have been
reported by HEINEN and HARTH (1976) and by GROTHE et al.
(1979) for diodes whose active layer is simply sandwiched
between a heterojunction and the contacted semiconductor surface. However, the efficiency of these extremely high frequency diodes is always relatively low.


## 6-4 Reliability

Today the reliability of LEDs is excellent. The lifetime is
usually defined as the time elapsed to reach a 50 % decrease
of the external efficiency. Good industrial products con-
stantly reach lifetimes above $10^5$ hours, or more than 11
years. After this time of continuous operation the diode
still works perfectly well, although with lower efficiency.

However, a quality this high has been achieved only after
tremendous efforts had been made towards identifying sources
of degradation, purification of materials and careful wafer
handling at all steps of processing. Moreover, sometimes insufficient knowledge of the interactions between wire bonds
and plastic encapsulation led to thermally induced bond in-
terruptions. Fortunately this is now a problem of the past.

Although the reliability of LEDs has been improved vastly
over the last decade, no single cause can be named for this.
A large bundle of influences can degrade LED performance and
is active most likely in varying combination depending not
only on the LED type but also on the fabrication laboratory.
In the following we describe characteristics and some specific mechanisms of degradation on an empirical basis. Finally
theoretical models are described which are able to explain
certain features of device degradation.

**6-4-1 Empirical characteristics of degradation**

In what follows twelve important, empirically found characteristic features of device degradation are described:

(1)   Degradation occurs only under forward bias. High
      temperature storage below 100° C or reverse bias
      does not influence device characteristics.

(2)   The degradation process slows down in course of
      time. The time dependence of the external efficien-
      cy frequently can be approximated by $\eta(t)/\eta(t_o) =$
      1-const log $t/t_o$. An example of this dependence is
      shown in Fig. 6-11.

(3)   The process of degradation is temperature activat-
      ed. Fig. 6-12 shows corresponding results with

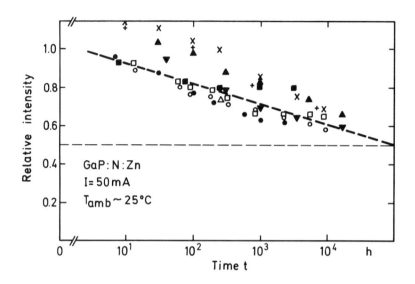

Fig. 6-11 Time dependence of the light output of LEDs
          from 8 different wafers stressed by a for-
          ward current of 50 mA, corresponding to a
          current density of 50 A/cm$^2$. Each point
          represents the average of 20 diodes, mount-
          ed on TO 18 headers without epoxy encapsu-
          lation. The pn junction temperature is
          estimated as 80° C and the lifetime extra-
          polates to at least $10^5$ hours.

activation energies of 0.53 eV and 0.76 eV. Temperature activation is usually used to derive room
temperature lifetime from accelerated tests at
higher temperatures. However, the corresponding law
of extrapolation can be perturbed when epoxy capsulated diodes are tested. This is because all epo-
xies get harder and shrink at low temperatures and
a hard encapsulant can have a detrimental influence

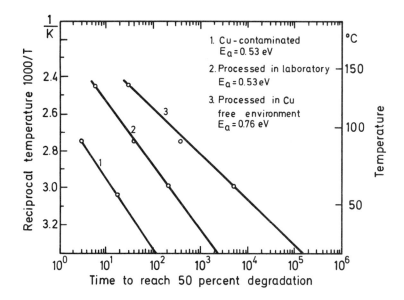

Fig. 6-12 Temperature dependence of diode lifetime.
        Plot of reciprocal temperature versus life-
        time of GaP:Zn:0 diodes, aged at $I_F$ =
        10 mA. Activation energies of 0.53 eV and
        0.76 eV can be derived for diodes either Cu
        contaminated (1 and 2) or processed in a
        Cu free environment (3). Data according to
        BERGH (1971).

on degradation through the mechanical stress it
exerts on the diode chip.

(4)  The current dependency of the degradation is often
     found to be superlinear with exponents of the cur-
     rent ranging between 1.5 and 2.0. A doubling of the
     forward current therefore means up to a fourfold
     increase of the degradation.

(5)  There is a tendency that materials emitting photons
     of low energy degrade slower than emitters of
     higher photon energy as was first pointed out by
     ETTENBERG and NUESE (1975). An explanation could be
     the recombination enhanced diffusion process de-
     scribed in section 6-4-3. Infrared emitters there-
     fore usually degrade slower than e.g. green emit-
     ting diodes.

(6)  The electrical forward characteristic deteriorates
     often during optical degradation. Fig. 6-13 shows
     the electrical characteristic of a diode before and

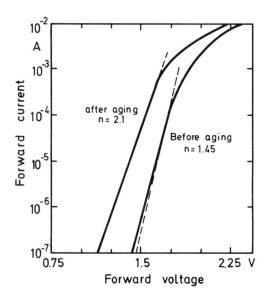

Fig. 6-13 Forward biased I-V characteristic of a
          diode before and after aging, according to
          BERGH (1971).

     after a lifetime test. The nonradiative n = 2 cur-
     rent component has distinctly increased, indicating
     an increase in space  charge recombination and thus

decreasing the injection efficiency of the pn junction. In a diode where this is the dominating degradation mechanism, the light output will be constant in time if the forward voltage is kept constant instead of the current.

(7) The lifetime $\tau$ and diffusion length L of injected
carriers has been often found to decrease during
optical degradation. Both are connected through L =
$\sqrt{D\tau}$ . In most diodes the nonradiative recombination dominates, i.e. $\tau_{nr} \ll \tau_r$. Therefore a reduction of the total lifetime means an increase in
nonradiative recombination.

(8) However, not only nonradiative recombination can
increase, but radiative recombination probability
can also decrease. This is caused by the rearrange-
ment or destruction of active recombination centers, as found by SCHAIRER (1979) as well as by
ALBRECHT (1983).

(9) The pn junction is not neccessarily operative in
evoking degradation processes. A decrease of the
optical efficiency of GaAs could be induced solely
by optical excitation as reported e.g. by MAHAJAN
(1979).

(10) During degradation frequently a rounding of the
electrical reverse characteristic is observed as
well as the formation of micro plasmas. Micro
plasmas are formed under reverse bias at spots of
high electric fields where an avalanche break
through happens before the uniform break through of
the junction. The avalanche break through creates
charge carriers - the plasma - whose white recombination radiation is easily recognized under a

microscope. The spots of high electrical fields are probably created by impurities accumulating at dislocations and forming spots of high doping levels. These effects have been described by SCHAIRER (1984).

(11) The n und p sides of the junction degrade often with differing velocity. This can produce a two step time dependence of the degradation process. An example in LPE GaP:N:Zn diodes has been analyzed by RHEINLÄNDER (1985).

(12) Finally one of the most prominent features of degradation in semiconductor lasers consists of dark

Fig. 6-14 $\{100\}$ emission face of a Zn-diffused GaAs diode with dark line defects parallel to $\langle 110 \rangle$ directions. The diode has been tested at 100 mA (1 kA/cm$^2$) for 1000 h, junction temperature about 100° C.

line defects (DLDs). In LEDs dark line defects
develop only at high current densities in high radiance emitters, used for glass fiber communication. Fig. 6-14 shows typical DLDs of a high radiance emitter.

## 6-4-2 Identification of some specific degradation mechanisms

By no means all degradation mechanisms are known and even
less have been investigated in some detail. Therefore only a
selection of active mechanisms can be presented here.

(1)   The kind of the bulk doping, be it intentional or
      not, can strongly influence the reliability of
      LEDs. Examples are the intentional doping of the
      n side of red emitting $GaAs_{0.6}P_{0.4}$ diodes by sele-
      nium, or the usually unintentional contamination of
      GaP diodes by copper. In the first case selenium
      was replaced by tellurium resulting in an improve-
      ment of the reliability by many orders of magni-
      tude. In the second case the very much improved
      cleanliness of the production environment eliminat-
      ed the reliability problem in accordance with the
      result shown in Fig. 6-12.

(2)   Surface contamination has also been identified as a
      source of problems. Spurious quantities of copper,
      or gold from alloyed contacts, apparently are able
      to enter the crystal under forward bias and can
      cause serious degradation of device performance.
      Fortunately it seems that not many chemical spe-
      cies, possibly only copper and gold, are able to
      enter the crystal sufficiently fast under normal
      operating conditions and are at the same time de-
      trimental to device performance.

(3)  The formation of nonradiative deep recombination
     centers has been observed in many types of diodes.
     The preferred detection technique is deep level
     transient spectroscopy (DLTS), a method that mea-
     sures changes of the junction capacitance connected
     with the thermal release of charges trapped at the
     centers to be detected. Although some of these
     centers are today quite familiar with respect to
     their energetical position, there exists little
     knowledge about their chemical and physical consi-
     tution. Therefore degradation caused by deep levels
     can be avoided presently only by the general ap-
     proach of high purity processing.

(4)  Mechanical stress is an important and rather ubi-
     quitous cause of degradation. Sources of stress are
     changes of lattice constant in mixed crystals, di-
     electrics on surfaces and most important sawing
     damage and plastic encapsulants. Detrimental ef-
     fects of the latter two are avoided by etching
     after sawing the LED dice and by using sufficiently
     soft encapsulants. Stress in LED chips can be made
     visible by placing the chip after removal of the
     reverse side contact between crossed polarizers.
     Stressed regions appear bright due to stress in-
     duced birefringence. Quantitatively the effect of
     stress has been studied by ZAESCHMAR and SPEER
     (1979) for GaAs IREDs. A threshold appears to exist
     above which strong degradation is provoked. The
     stress induced degradation is explained by the
     movement and multiplication of dislocations. The
     mobility of dislocations is strongly enhanced by
     the (nonradiative) recombination of charge carriers
     as pointed out by MAHAJAN et al. (1979).

## 6-4-3 Theoretical models

The problem to be explained is the occurrence of degradation
processes at junction temperatures below 100° C. Purely thermally activated diffusion for example is much too slow to
explain short diode lives. This view is empirically confirmed
as described in point (1), section 6-4-1.

HENKEL (1962), and also LONGINI (1962), have proposed that
ionized impurities, similar to electrons and holes, can cross
the pn junction when the built-in voltage is reduced through
a forward bias. The solubility of these impurities could be
reduced on the respective other side of the junction because
of the shifted Fermi level, thus favouring precipitation and
the formation of nonradiative centers. Most likely, positive-
ly charged interstitital Zn could be such an impurity. This
model explains why degradation occurs only under forward
bias, but it cannot explain the very low temperature diffusion of the ionized impurities, much less e.g. an observed
purely optically induced degradation; see point (9), section
6-4-1.

GOLD and WEISBERG (1964) proposed that an impurity, e.g. the
intentionally introduced Zn, can be displaced to an interstitial site through the energy provided by nonradiative recombination. The resulting complex, a vacancy-interstitial ion,
could be directly responsible for nonradiative recombination
or indirectly lead to nonradiative centers. Similary, the intense lattice vibrations caused by nonradiative recombination
could enhance diffusion processes at very low temperatures.

These ideas have been substantiated and quantitatively tested
by KIMERLING and LANG (1975) in recombination enhanced annealing experiments. They introduced 0.31 eV electron traps
into n type GaAs by electron irradiation. Thermally annealed,

these traps disappear, yielding an activation energy of
1.4 eV, as shown in Fig. 6-15 according to KIMERLING and
LANG's work. The activation energy decreases to 0.34 eV, if
the annealing experiment is repeated under sufficiently high
forward bias. The difference between the two activation
energies of 1.06 eV cor-
responds to the energy
released nonradiatively
when a trapped electron
recombines with a free
hole. The disappearence
of the trap implies that
defects move in the cry-
stal more readily under
forward bias than without
it. This effect is called
recombination enhanced
diffusion. It is believed
that the mechanism de-
scribed here, studied for
a very specific center,
is quite generally active
in degrading diodes. This
effect explains where the
energy needed to induce

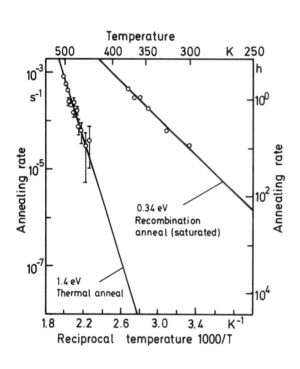

Fig. 6-15 Annealing rates of a 0.31 eV electron trap in GaAs
         diodes as a function of reciprocal temperature.
         Forward bias strongly increases the annealing
         rates: recombination anneal. The difference E be-
         tween the corresponding activation energies E =
         1.4 eV - 0.34 eV = 1.06 eV matches the Energy E'
         released during the recombination of a trapped
         electron with a hole of the valance band: $E' = E_g$
         - 0.31 eV = 1.4 eV - 0.31 eV = 1.09 eV.

defect reactions comes from. The major part is provided by
nonradiative recombination; only a minor part is thermal
energy. The model also explains that lower band gap materials
tend to degrade slower than higher band gap materials; see
point (5), section 6-4-1.

The energy released during nonradiative recombination can
also cause movements of dislocations via defect reactions, as
shown by KIMERLING et al. (1976).

The growth of dislocation lines proceeds through glide or
climb processes depending on the material and the crystal
orientation. The speed of the growth depends on the rate of
nonradiative recombination, but increases also with temperature and mechanical stress. Temperature and stress are operative over the whole diode chip, whereas carriers recombine
nonradiatively only in a thin layer accompanying the junction. Therefore dislocation lines originally crossing the
junction on the shortest way tend to form extended dislocation loops within the recombination zone. The loops grow not
statistically, but follow "easy" crystal plains and create
therefore characteristically oriented DLDs. The increased
density of dislocation lines in the active region of diodes
lowers their internal efficiency seriously. The formation of
DLDs can be completely prevented by selecting material which
has a dislocation density low enough to have not a single
dislocation line running through the active zone of the
diode.

## Questions

Q6-1    Assume that the radiant flux P of a diode is propor-
        tional to the diffusion current (Eq. 6-8) and that the
        current due to space charge recombination (Eq. 6-9) is

large compared with the diffusion current. Show that the radiant flux is proportional to the square of the forward current.

Q6-2    Assume the validity of Eq. (6-10) and (6-11) for internal efficiency and recombination rates. Express the internal efficiency and the total lifetime through radiative and nonradiative lifetimes according to Eq. (6-12) and (6-13).

Q6-3    Calculate the angle of total reflection for a GaAs ($n_1$ = 3.6) air interface using Eq. (6-28). Do it again, replacing air by epoxy ($n_2$ = 1.5).

Q6-4    Estimate which fraction f of the power of an isotropic radiation source inside GaAs can penetrate the surface in a first attempt. Use the results of Q6-3 and Eq. (6-30).

Q6-5    Differing from the text, a diode's modulation bandwidth can be defined as the frequency at which the modulation efficiency has dropped to 50 % (- 3 dB) of its static value. Calculate accordingly the bandwidth of a diode whose frequency response is described by Eq. (6-39).

## References

ALBRECHT, H., "The Role of N on the degradation behavior in LPE and VPE GaP:N LEDs," IEEE Trans. Electron. Dev., ED-30 (1983), pp. 259-263

BEBB, H. B. and E.W. WILLIAMS, "Photoluminescence I: Theory," Semiconductors and Semimetals, Vol. 8, Ed. by R.K. Willardson and Albert C. Beer, Academic Press (1972), New York and London

BERGH, A. A., "Bulk degradation of GaP red LEDs," IEEE Trans.
    Electron Dev., ED-18 (1971), pp. 166-170

ETTENBERG, M. and C.J. NUESE, "Reduced degradation in
    $In_xGa_{1-x}As$ electroluminescent diodes," J. Appl. Phys.,
    46 (1975), pp. 2137-2142

GILLESSEN, K. and A.J. MARSHALL, "Correlation between loca-
    tion of light emission and spatial N distribution in GaP
    LEDs," IEEE Trans. El. Dev., ED-26 (1979), pp. 1186-1189

GOLD, R. D. and L.R. WEISBERG, "Permanent degradation of GaAs
    tunnel diodes," Solid-State Electron., 7 (1964), pp.
    811-821.

GROTHE, H., W. PROEBSTER and W. HARTH, Electron. Lett., 15
    (1979), pp. 702-703

HART, P.B., "Green and yellow emitting devices in vapor-grown
    GaP," Proc. IEEE, 61 (1973), pp. 880-884

HEINEN, J. and W. HARTH, "Light-emitting diodes with a modu-
    lation bandwidth of more than 1 GHz," Electron. Lett.,
    12 (1976), pp. 553-554

HENKEL, H.-J., "Alterungserscheinungen an GaAs-Tunneldioden,"
    Z. Naturforschg., 17a (1962), pp. 358-360

JOYCE, W.B., R.Z. BACHRACH, R.W. DIXON and D.A. SEALER, "Geo-
    metrical properties of random particles and the extrac-
    tion of photons from electroluminescent diodes," J.
    Appl. Phys., 45 (1974), pp. 2229-2253

KIMERLING L.C. and D.V. LANG, "Recombination enhanced defect
    reactions in semiconductors," Proc. Int. Conf. on the
    Lattice Defects in Semiconductors, Freiburg, 1974, Inst.
    Phys. Conf. Ser., 23 (1975), pp. 589-593

KIMERLING, L.C., P. PETROFF and H.J. LEAMY, "Injection-stimu-
    lated dislocation motion in semiconductors," Appl. Phys.
    Lett., 28 (1976), pp. 297-300

LONGINI, R.L., "Rapid Zn diffusion in GaAs," Solid-State
    Electron., 5 (1962), pp. 127-130

MABBIT, A.W. and R.C. GOODFELLOW, "High-radiance small-area
    GaInAs 1.06 μm light-emitting diodes," Electron. Lett.,
    11 (1975), pp. 274-275

MAHAJAN, S., W.D. JOHNSTON, Jr., M.A. POLLACK and R.E.
    NAHORY, "The mechanism of optically induced degradation
    in InP/In$_{1-x}$Ga$_x$As$_y$P$_{1-y}$ heterostructures," Appl. Phys.
    Lett., 34 (1979), pp. 717-719

PILKUHN, M.H., "Light emitting diodes," Handbook on Semicon-
    ductors, ed. by T.S. Moss, Vol. 4, Chapter 5A, North
    Holland Publishing Co. (1981)

REICHL, H., J. MÜLLER and D. HUBER "A method for the deter-
    mination of optimum p-n junction depth of luminescence
    devices," J. Appl. Phys., 49 (1978), pp. 4838-4842

RHEINLÄNDER, B., G. OELGART, H. HAEFNER and R. PICKENHAIN,
    "Degradation of green LEDs LPE-GaP:N, II.," Phys. stat.
    sol., (a) 87 (1985), pp. 373-381

ROOSBROECK, W. van and W. SHOCKLEY, "Photon-radiative recombination of electrons and holes in Ge," Phys. Rev., <u>94</u> (1954), pp. 1558-1560

SCHAIRER, W., "Defect centers and degradation of GaP:N LEDs," J. Electron. Mater., <u>8</u> (1979), 139-151

SCHAIRER, W., "Rapid thermal redistribution of Cu impurities in $GaAs_xP_{1-x}$ at temperatures above 250° C," J. Electron. Mater., <u>13</u> (1984), 559-574

ZAESCHMAR, G. and R.S. SPEER, "Mechanical-stress-induced degradation in homojunction GaAs LEDs," J. Appl. Phys., <u>50</u> (1979), 5686-5690

# 7 APPLICATIONS

## 7-1 General aspects

Because light emitting diodes are solid state, semiconductor
light sources, they have some specific properties in common
with other semiconductor devices like transistors or integrated circuits. Some general features of LEDs and other
semiconductor devices are small size, low weight, high mechanical stability, low temperature sensitivity, high reliability, long operating lifetime, and last but not least, low
price. As electrical devices, LEDs are characterized by low
operating voltage, medium current, and high speed. From an
optical point of view, the most important properties of LEDs
can be summarized as follows: LEDs are active emitters of
nearly monochromatic light with highly saturated colors.

From this set of properties some specific fields of application can be derived, which are treated in detail in the following parts of this chapter. Only a few general remarks will
be made here. LEDs are mostly used at the periphery of electronic equipment of all kind. Interfacing LEDs to electronics
is especially easy, because their driving requirements can be
easily fulfilled by standard transistors and integrated circuits. This property is decisive for the bulk of LED applications. Typical examples are status indicators, displays for
entertainment electronic equipment, and displays for measuring equipment. LEDs are also preferred to other display tech-

nologies in rough environments and where high reliability is
imperative.

On the other hand, the properties of LEDs define also the
limits of their applicability. For example, LEDs are not well
suited for general illumination purposes, because their
brightness is still inferior to other light sources, and because white light can hardly be achieved (blue LEDs are less
efficient than red, yellow and green devices), but is required for color-neutral illumination. LEDs are also rarely used
in battery operated equipment, where power consumption is a
critical factor. Here liquid crystal displays (LCD) are dominating. The relatively high power consumption, and consequently, power dissipation of LEDs renders also the realization
of high resolution, flat LED screens more difficult (see
section 7-3-4). For color picture displaying, the inferior
brightness of blue emitters is another limitation. Other competing flat screen technologies like LCDs or plasma displays
also have difficulty in displacing the cathode ray tube,
which is well developed. It offers rather high performance at
quite low cost.


## 7-2 Driving of LEDs

### 7-2-1 Current limiting

The I-V characteristics of LEDs are those of normal pn
diodes: in the forward direction, which is the normal operating mode, the current remains rather small until the voltage
amounts to approximately $E_g/e$, which is about 1.9 V for red
LEDs, 2.0 V for orange, 2.1 V for yellow, and 2.2 V for
green. Around these voltages the current rises exponentially
(see chapter 1). Because the brightness of an LED is determined by the current flowing through the device, it is the

current which has to be defined during operation. If an LED
is driven by a voltage source, its brightness is very sensitive to small voltage fluctuations, due to the steep I-V characteristics. Therefore, some means for current limitation
has to be provided. Two examples for current definition in
LED circuits are shown in Fig. 7-1. The simplest possibility
is a resistor in series to the LED (Fig. 7-1, left). Here the
current through the LED ($I_{LED}$) is defined by the intersection
of the LED characteristics with the straight line given by
(operating voltage minus LED voltage) divided by resistance,
which is (5 volts-$V_{LED}$)/150 $\Omega$ in the example shown. As can be
seen easily, the LED current is now far less sensitive
against voltage changes. In most practical applications the
value of the resistor can be calculated using the simple formula:

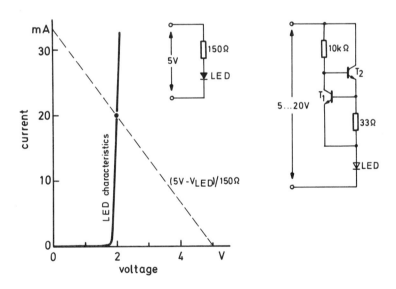

Fig. 7-1 Methods of control of LED current, left: LED
         with series resistor, determination of work-
         ing point, right: LED with constant current
         source for approximately 20 mA.

$$R = \text{(operating voltage - 2 V)/LED current} \qquad (7\text{-}1)$$

because the forward voltage required for a normal operating
current is roughly 2 volts for all types of LEDs.

If large voltage variations are to be expected, LEDs are
driven best using a constant current source. For this purpose
integrated circuits are available which are supplied by several manufacturers. A simple constant current source can also
be realized with two transistors and two resistors as shown
on the right hand of Fig. 7-1. This circuit controls the LED
current via the voltage drop along the 33 $\Omega$ resistor, which
is 0.66 V at 20 mA output current. If the current increases,
the transistor $T_1$ becomes more conducting, thus diminishing
the current flowing into the base of the transistor $T_2$.
Therefore, the output current is decreased again. The value
of the second resistor is determined by the minimum voltage
to be expected and the base current necessary to drive $T_2$. In
the example shown it is assumed that at least 2 V are available at this resistor (5 V minus LED voltage minus collector
emitter voltage of $T_2$), and that 0.2 mA are sufficient to
cause an output current of 20 mA, i.e. the current gain of $T_2$
should be at least 100. The upper voltage limit is given by
the power dissipation in $T_2$ which is roughly 20 V times 20 mA
or 0.4 W.

## 7-2-2 Multiplex operation

Because the light output of an LED is proportional to the
operating current over a wide current range, LEDs can be
driven by pulses instead of continuous current. The apparent
brightness is then given by the average current. Some LED
types even exhibit a superlinear current-light relationship,
so that a net gain in brightness results in pulsed operation.
This property is widely used in time multiplex operation of

LED displays. A simple example consisting of a four-digit, seven-segment display is depicted in Fig. 7-2. The equivalent segments of all four digits are connected to one segment line, consequently there are seven segment lines. In addition, four common lines are provided, one for each digit. The driving circuitry necessary for operation of this display is indicated by the switches in the lines. The status sequence for displaying "1 2 3 4" is shown in the bottom part of Fig. 7-2.

Only one digit is activated at one time, by cyclic closing of the switches in the common lines C1 to C4 (three cycles are shown). To display a particular numeral, current is supplied to the corresponding segment lines, i.e. to S2 and S3 for "1", to S1, S2, S4, S5, S7 for "2" and so on. To achieve the same brightness as with continuous current operation, each LED is driven with a four times larger current for a quarter of the time in this example. The minimum frequency required for continuous appearance of the display is about 30 cycles per second. This

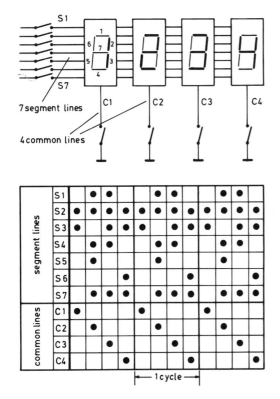

Fig. 7-2. Multiplex operation of LED display, top: configuration, bottom: addressing scheme

frequency is given by the properties of the human eye.

### 7-2-3 Interconnection schemes

In LED displays containing several diodes it must be possible
to address every LED separately. The most straightforward way
to achieve this is to provide one separate line for each LED
plus one common line. For displays with more than about 10
LEDs the large number of adress lines can result in problems,
due either to the large number of output stages necessary, or
to the space required for lines and connectors. It is therefore worthwhile to look for methods to reduce the number of
address lines. The most common method used for this purpose
is to interconnect the LEDs in the form of a matrix. This was
already implicitly shown in the four-digit display in Fig.
7-2, where 4 x 7 = 28 LEDs were addressed via 7 row and 4
column lines, i.e. with a total of 11 lines. Among different
matrix configurations the most favourable ratio of LEDs per
lines is achieved with a square matrix, i.e. with an equal
number of rows and columns. If a number n of address lines is
assumed, n/2 rows and columns each can be formed, which results in a number of $(n/2)^2 = n^2/4$ addressable LEDs. Separate
addressing of LEDs with a matrix-like interconnection scheme
is possible only because LEDs are also diodes in an electrical sense, featuring reverse blocking characteristics.

The use of the electrical properties can be extended even
more, resulting in a further increase of the number of addressable LEDs. Two further interconnection schemes are depicted in Fig. 7-3. For comparison the standard matrix is
also shown (Fig. 7-3, top). The two other interconnection
methods rely on the reverse blocking as well as on the forward threshold characteristics of LEDs. If one diode is connected in parallel to two diodes in series, a noticeable current flows only through the single diode because of the

strongly nonlinear current-voltage relationship of LEDs. It
is therefore possible to connect one pair of antiparallel
diodes to all combinations of two address lines. This connection method is therefore called "combinatorial". It can be
shown that $n(n-1)$ diodes can be driven with only n lines
using this combinatorial principle. In the example shown in
Fig. 7-3, middle, the LEDs have been rearranged to form a
nearly square matrix with
n rows and n-1 columns.

If some of the antiparal-
lel diode pairs are sa-
crificed, the remaining
LEDs can be interconnect-
ed in a cross-free man-
ner. This is a consider-
able advantage, because
interconnection can then
be realized for example
with a single layer me-
tallization of a PC board
or with a similar simple
technique. In this case,
4n-6 diodes can be sepa-
rately addressed with n
lines. Fig. 7-3, bottom,
gives an example for this

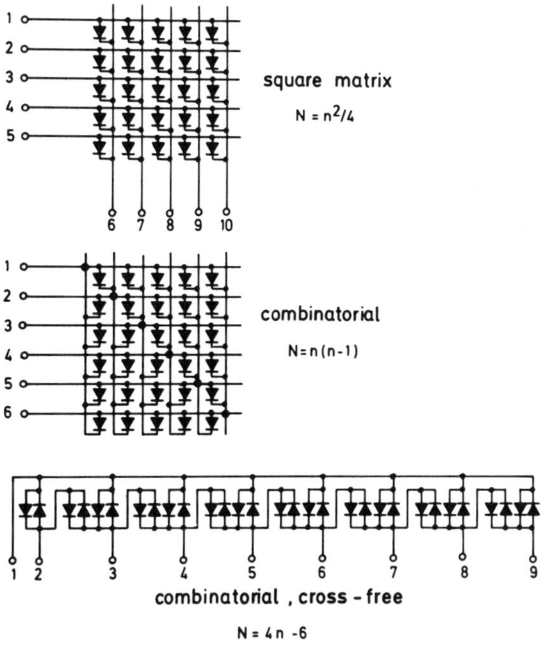

square matrix

$N = n^2/4$

combinatorial

$N = n(n-1)$

combinatorial , cross - free

$N = 4n - 6$

Fig. 7-3 LED interconnection schemes for reduction of the
number of address lines, N = LED number, n = line
number. top: square matrix with 25 LEDs and 10
lines, $N = n^2/4$, middle: combinatorial interconnection with 30 LEDs and 6 lines, $N = n(n-1)$, bottom:
combinatorial, cross-free interconnection with 30
LEDs and 9 lines, $N = 4n - 6$

combinatorial, cross-free interconnection method, with the
LEDs forming a one-dimensional array in this particular case.
To compare the interconnection schemes, the respective numbers of LEDs addressable with n lines are plotted in Fig.
7-4, which also includes the straightforward method of addressing by separate lines. In the case of the square matrix
the LED number increases quadratically with the number of
lines. Using the combinatorial principle, about four times
more LEDs can be addressed than with the matrix. The cross-
free interconnection method is identical with the combinato-

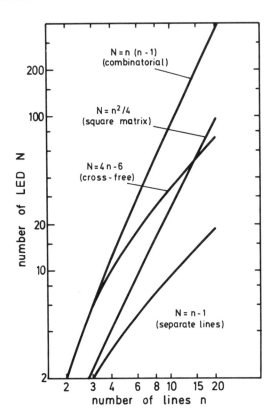

rial one for n = 2 and
n = 3 and allows addres-
sing of a larger number
of LEDs than the matrix
method up to n = 14,
where the respective LED
numbers are 50 (cross-
free) and 49 (matrix).
For larger n, approxima-
tely 4 LEDs per line can
be driven.

The driving circuitry for
LED displays is apparent-
ly quite complicated, so
that it can be realized
economically only by
using integrated cir-
cuits. At present appro-
priate ICs are available

Fig. 7-4 Comparison of numbers of LEDs for different inter-
         connection schemes: N = n-1 (separate lines), N =
         4n-6 (cross-free), N = $n^2$/4 (square matrix), N =
         n(n-1) (combinatorial)

for matrix and cross-free interconnected LED displays.

## 7-3 Application of visible LEDs

Visible light emitting diodes are used in large numbers in
many fields of application, and of course we cannot mention
all imaginable employments. In most cases the detector of the
light of LEDs is the human eye. We have arranged these applications in the order of increasing complexity, starting with
single indicator lamps, then numeric and alphanumeric displays, one-dimensional arrays, and finally flat screens. The
paragraph on further applications includes use of LEDs in

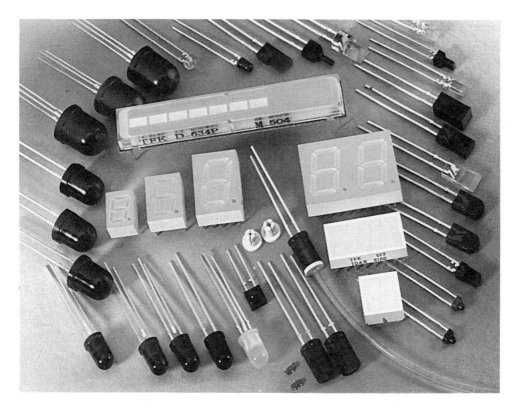

Fig. 7-5  Typical applications of visible LEDs
(Photograph courtesy by TELEFUNKEN electronic GmbH)

conjunction with other detectors and an example for non-con-
ventional use of LEDs. Some typical LED applications are
shown in Fig. 7-5, including single LEDs in different sizes,
colors, and shapes, special LEDs, some numeric displays, and
a linear array of LEDs used as audio level indicator.

### 7-3-1 Indicator lamps

The simplest application of LEDs is the use of single LEDs as
status indicators, visualizing for example an "ON" condition
or a failure of an apparatus. Different LED colors can be
used to distinguish between different functions to be displayed. For these standard purposes LEDs are mostly packaged
in plastic cases, where the epoxy material serves simultaneously as mechanical fixture, encapsulation, and optical
element, e.g. light conductor, lens, or diffusor. Three cus-

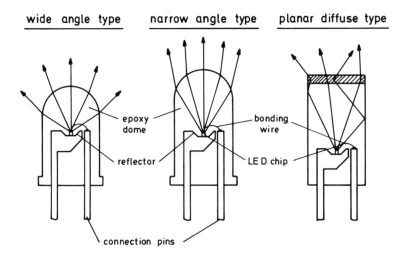

Fig. 7-6   Single LEDs in plastic cases, left: wide
           angle type, center: narrow angle type,
           right: planar diffuse type

tomary types of plastic cases are shown in Fig. 7-6. The LED
chip is fixed in a recess of a metal lead frame which later
forms the connection pins. The recess serves as a reflector
which directs light from the side faces of the chip towards
the axis of the LED case. In most LEDs the chip is fixed
using a conducting adhesive, usually an epoxy resin filled
with silver powder. Only in high power types is the chip sol-
dered to the metal base to achieve a very low thermal resistance. The top contact of the LED is then connected to the
second pin with a wire bond, most frequently using the so
called thermocompression process with gold wire. Thermocompression means that the wire is pressed onto the contact at
somewhat elevated temperature (250 to 350° C), forming a kind
of weld connection. Instead of or in addition to high temperature ultrasonic energy can also be used to form the connection. These processes are named ultrasonic and thermosonic
bonding, respectively. Using a casting process, the lead
frame is then embedded in an epoxy body, the color and shape
of which depend on the assigned application. Finally, the
lead frame is cut into separate pins. The emission angle of
standard round LEDs can be adjusted by variation of the distance between the chip position and the epoxy dome acting as
a lens (Fig. 7-6, left and middle). The epoxy material can be
clear or diffuse, and colorless or tinted. In the latter case
the emission color is hardly changed, instead the coloration
renders it possible to discern LEDs of different colors without actual operation. With the third type (Fig. 7-6, right)
the epoxy body consists of two parts: a clear one which serves as a light conductor, and a diffuse part with a planar
homogeneously radiating surface. This LED type is produced
with a variety of cross sections, so that the surface can
have many different shapes (circular, square, rectangular,
triangular, and so on). With these LEDs simple symbols (like
arrows) can be incorporated in a front panel design, or the
difference in shape can be used as a second distinctive mark

besides the emission color. Larger planar diffuse LEDs may contain several LED chips to achieve the required brightness. They are used for reverse side illumination of symbols and short messages to the operator of an apparatus.

Apart from the standard devices there are also LEDs with some additional features available. For example, in order to simplify the use of LEDs, frequently used circuit elements can be incorporated in the same package as the LED chip. In the simplest case, this is a series resistor for current limiting as described in section 7-2. For operation from varying voltages, LEDs with built-in constant current sources are offered. For indicators which require special attention, there are blinking LEDs in various versions which contain an integrated circuit which switches the LED on and off with a typical frequency of 2 Hz. Blinking LEDs with a third (control) terminal can be operated either blinking or with continuous light. The special LEDs also include two-color types with, for example, one red and one green chip in the same package. This makes it possible to generate three colors by activating either the red or the green or both chips simultaneously. In the latter case, yellow light results from additive mixing of red and green. In two-color LEDs, the two chips are either connected in antiparallel or a third terminal is provided which, for example, is connected to both of the cathodes. With these two-color LEDs different operating states, transgression of limit values or similar conditions can be visualized particulary easily and obviously.

## 7-3-2 Numeric and alphanumeric displays

For representation of the ten numerals "0" to "9" the "8"-shaped seven-segment display is widely used, as shown in Fig. 7-7, top row. With 7 or 8 LEDs, including a decimal point, the numerals can be formed with reasonable legibility. Capi-

tals and the numerals can be displayed using a 14-segment,
so-called flag display, see Fig. 7-7, middle part, although
the representation of some capitals is not particularly
pleasing, for example "B", "D", "J", and "Q". With a matrix
of 5 x 7 dots numerals, large and small characters, and some
special characters can be formed in good quality, as depicted
in Fig. 7-7, bottom part.

| display-type | number of LEDs | design | examples |
|---|---|---|---|
| numeric | 7 | 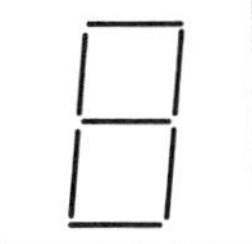 |  |
| flag | 14 | 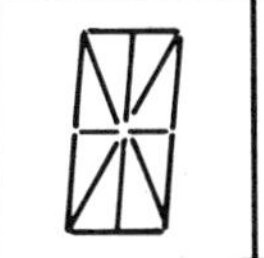 | 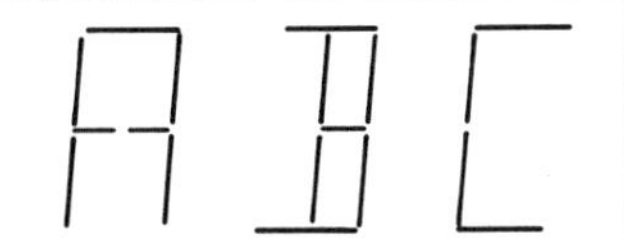 |
| 5 x 7 matrix | 35 |  | 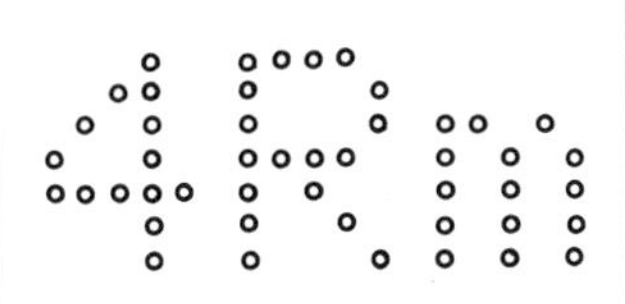 |

Fig. 7-7 Numeric and alphanumeric displays, top: se-
         ven segment numeric display, middle: 14-seg-
         ment flag display, botton: 5 x 7 matrix dis-
         play

The majority of LED chips is used in displays of this kind.
Well-known fields of application are entertainment electronic
equipment like radios, TV sets, and video cassette recorders,
measuring equipment like voltmeters, thermometers, and balan-
ces, calculators and clocks. In small battery operated equipment like pocket calculators and in particular wrist watches
LED displays were however superseded by liquid crystal displays (LCDs) which consume much less power.

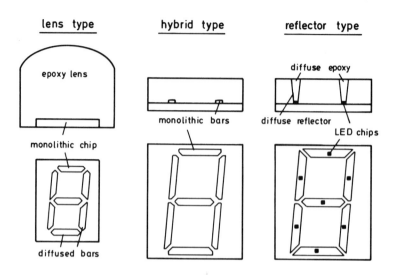

Fig. 7-8 LED display types, left: monolithic, lens
         type, middle: hybrid type, right: reflector
         type

There are several possible ways of constructing LED displays
economically and with pleasing appearance, three of which are
shown in Fig. 7-8. For small, low cost displays the monolithic, lens type display is preferred (Fig. 7-8, left). The
required pattern is formed by diffusion of separate p regions
into an n-type monolithic chip. Therefore, the resulting LEDs
have a common cathode. The image formed by the chip is enlarged by the lens molded from clear epoxy. Slightly larger
displays can be composed in a hybrid technique, where monolithic LED bars are fixed on a substrate and covered with
clear plastic (Fig. 7-8 middle). Most LED displays are however of the reflector type shown in Fig. 7-8 right. Depending
on the size, one or several chips are provided per segment.
The chips are placed within cavities in a reflector which
consists of white plastic with a diffuse reflecting surface.

The cavities are then filled with transparent diffuse epoxy,
so that the chip is not directly visible. Instead, the light
emitted from the chip is homogeneously distributed over the
respective segment. Reflector type LED displays are available
with common cathode or anode, in different colors, and in
size ranging from 7 to more than 25 mm character height.
The same tendency as with single LEDs can also be observed
with LED displays: they are integrated into larger units,
which also may include decoding and driving circuits, forming
so-called intelligent displays. With such units, the user is
provided with functional blocks which can be applied very
easily.

### 7-3-3 One-dimensional arrays

Another important field of LED application is the formation
of one-dimensional arrays. These are fabricated for various
purposes in strongly varying sizes ranging from 5 to several
thousand LEDs. The smallest versions are volume indicators in
audio equipment, which contain 5 to 10 LEDs plus the required
driving circuits (see also Fig. 7-5). Medium-sized arrays
with 16 to 150 LEDs can be used as tuning scales, level indicators, or generally as quasi-analog displays, which can
replace pointer instruments in many cases. Such quasi-analog
displays can be implemented as point indicators (flying dot
displays) or bar-graph displays (thermometer type display).
One-dimensional arrays with even larger numbers of LEDs and
consequently higher resolution can be used for representation
of pictures. For this purpose the array displays the information of one row (or column) of the picture at a time, while
it is scanned over the field of view. Three examples for the
application of this principle will be given here. The first
one is a display used in a night-vision system, which allows
observation of infrared radiating objects in the dark, a
task which is important for both military and civilian use,

e.g. detection of a tank or of insufficiently insulated parts
of a house. The picture is taken up by a infrared-detector
arrangement, which is sensitive, for example, in the 5 to
10 µm region. The signal is then amplified and given to an
LED array with e.g. 256 elements on a 16 mm line. This array
is observed with an optical system including a rotating mirror, which generates the appearance of a two-dimensional picture.

The second example for picture generation using one-dimensional LED arrays is described by NAKAYA et al (1982). The LED
array consists of 16 monolithic chips with 128 LEDs each at
intervals of 125 µm. The total number of LEDs is therefore
2048, and the array length amounts to 256 mm. By observation
over a rotating mirror the apparent number of picture elements is 2048 x 360 = 737280.

In our third example an LED array is used as an exposure module in a xerographic printer. The xerographic principle is
well-known from copying machines: a rotating drum covered
with a photoconductive material like selenium is charged and
exposed with the image to be printed. The exposed parts of
the surface are discharged by photoconduction, while the dark
parts remain charged. A toner is then applied, which adheres
only at the charged parts, and is transferred to the paper,
where it forms a black and white picture. If this principle
is to be applied to a printer, the information to be printed
is usually written linewise to the surface of the photoconductive drum, whereas in copying machines the picture as a
whole is projected onto the drum with an optical system. For
linewise exposure three methods can be used: (1) scanning
with a modulated laser beam, (2) illumination over a light
shutter array, and (3) direct exposure by a high-resolution
LED array. In the following we will describe an LED array,
which was developed for the last method.

The LED array consists of 20 monolithic planar diffused
$GaAs_{1-x}P_x$ chips with 128 LEDs each, i.e. it comprises a total
of 2560 LEDs (see Fig. 7-9 and Fig. 7-10). The chips are
mounted on a ceramic substrate with high accuracy, so that
the center-to-center distance of 85 µm is maintained over the

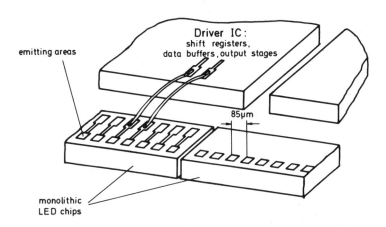

Fig. 7-9 LED array for printer applications (schematically) only a small section of the array
is shown, which actually consists of 2560
LEDs

whole length of the array. This pitch corresponds to a resolution of about 12 points per mm. Driver ICs are also located
on the same substrate, which contain shift registers, data
buffers, and output stages. A single element of this array
emits about 10 µW of red light at a typical driving current
of 3 mA. The main advantages of this LED array over competing
exposure methods are mechanical stability, high reliability,
low power consumption, small space requirements, and simplification of printer construction, because there are no moving
parts in the exposure module.

Similarly to other LED applications, one-dimensional arrays
also tend to be integrated with the required driving circui-
ry, to provide easily applicable units.

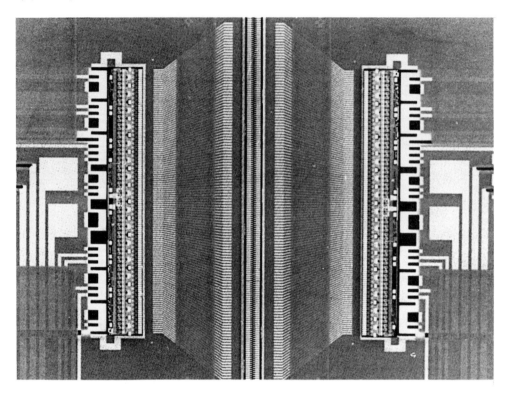

Fig. 7-10 Microphotograph of a section of the LED
          array for printer applications. The LEDs
          are located on a vertical line in the cen-
          ter of the figure, with driving circuits
          placed on both sides.
(Photograph courtesy by TELEFUNKEN electronic GmbH)

### 7-3-4 Flat screens

When the first light emitting diodes were introduced, it was
the general opinion that flat screens in LED technology could
soon replace the bulky cathode ray tube for example in TV
sets. The actual development has shown, however, that the CRT

was very much reduced in depth; even "flat" tubes with the
electron beam nearly parallel to the screen over most of its
path were constructed for use in pocket TV receivers. Other
flat screen technologies were also developed, for example
plasma displays and liquid crystal displays, and TV sets with
LCD screens are available today. Flat screens based on LEDs
have up to now been realized only on an experimental scale.
We will describe some of these experimental systems in this
section, and point out the limitations of the present LED
technology with respect to flat screens.

It is of course possible to construct flat screens by assem-
bling a large number of individual LED chips on a common carrier. The spatial resolution of such an assembly is however
limited by the chip size of typically 0.3 mm, which cannot be
reduced very much further because handling of even smaller
chips is very difficult. On the other hand, the large number
of chips required for acceptable picture quality results in
an equally large number of wire bond connections. The assem-
bling of an LED display consisting of individual chips is
therefore a formidable task which precludes extended use of
this technique. Displays of this type are for example described in the reference by NIINA et al. (1979). The largest
device comprises 38400 LEDs on a 160 mm x 120 mm area, which
are interconnected according to a matrix scheme with metal
stripes on the substrate forming the row connections, and
bonding wires leading from one chip anode contact to the
other serving as column lines. With a driving current of
0.5 mA per dot a total brightness of 137 cd is achieved. The
maximum total driving current can be calculated to be 19.2 A,
resulting in a power dissipation of more than 40 W for the
LED display alone without periphery.

By analogy to silicon technology monolithic integration
should obviously be employed to reduce the size of individual

LEDs and to rationalize the fabrication of the interconnection lines. The technology of the LED materials is however
far less advanced than silicon technology. In particular, n
diffusion and the formation of buried layers cannot be easily
performed in III-V compounds. Second, the planar LED technology based on vapor phase epitaxy and Zn diffusion, which is
better suited for monolithic integration, is inferior with
respect to brightness to the liquid phase epitaxial technology, which renders separation of individual diodes more difficult. In addition, monolithic displays require not only
electrical but also optical isolation of the single elements
to achieve sufficient contrast between ON and OFF elements.
These factors have hindered the development of monolithic LED
screens. As an example for the state of the art the reader is
referred to the paper of YODOSHI et al. (1984). It describes
fabrication and properties of a monolithic two-dimensional
array of 3600 GaP LEDs. There are two layers of metallization
for row and column lines, connected to the anodes and cathodes, respectively, of the individual diodes. Electrical and
optical isolation of the elements is achieved by a double
mesa etching process, with the grooves being refilled with
polyimide resin. The total power consumption of the 20 mm x
20 mm chip is only 300 mW. Internal light reflection within
the chip limits the contrast ratio to about 10:1.

As can be seen from the examples just mentioned, flat screens
on the basis of light emitting diodes still have to be very
much improved before they can really compete with other display technologies. The most important obstacles hindering
faster progress in development are limited brightness or high
power consumption, inferior brightness of blue LEDs, and difficulties in monolithic integration.

### 7-3-5 Further applications

Visible light emitting diodes are also used in areas where
their radiation is detected not by the human eye but by some
other device, i.e. where their visibility is irrelevant. Such
applications are closely related to the field of applications
of infrared emitting diodes, which is described in section
7-4. Therefore the use of visible LEDs is treated here only
briefly.

LEDs can be used in couplers which are sometimes also called
opto-isolators. Although the power of LEDs is much smaller
than that of IR devices (see chapter 6), they are generally
faster than infrared emitters. Red LEDs on the basis of
$Ga_{1-x}Al_xAs$ feature a particularly favourable combination of
high speed and rather high power at low cost, so they are
employed in fast couplers.

Visible LEDs are also advantageous over IR emitters in conjunction with integrated silicon detectors. The structure of
a discrete detector can be optimized with respect to sensitivity and speed. This degree of freedom does not exist for
integrated silicon detectors (detectors with integrated pre-
amplifiers, logic circuits and so on), where the choice of
layer thicknesses and doping levels is restricted by technological requirements form the electrical parts of the circuits. When excited by infrared radiation, integrated detectors tend to be slow, because the carriers are generated in a
larger volume and have to diffuse long paths to the sensitive
pn junction. Again, red $Ga_{1-x}Al_xAs$ LEDs offer a good compro-
mise with respect to speed, power, and penetration depth into
silicon.

The same devices are also applied for data transmission over
short distance via inexpensive plastic fibers. Compared to

optimized systems employing quartz fibers and specialized devices like edge emitters or lasers, rather low data rates of several Mbit/s can only be transmitted with such low cost systems. There are, however, many applications, like in-house connections or computer periphery links, the requirement of which can be fully met with simple systems employing standard LEDs.

Furthermore, LEDs are used in increasing numbers in optical switches for industrial applications, like determination of the position of an object on an assembly line. Here the visibility is advantageous because it simplifies adjustment and function check of the system.

Finally, light emitting diodes are also of practical use as purely electrical devices regardless of the emitted radiation. The forward voltage drop of 1.6 V to 2.2 V is considerably higher than that of silicon diodes (0.7 V) and can be applied to shift a potential in a circuit. The steep rise of the forward characteristics can also be used for voltage stabilization. In fact, the differential resistance (which determines the performance of a reference diode) of forward biased LEDs is smaller than that of silicon Zener diodes in the same voltage range.

## 7-4 Application of infrared emitters

### 7-4-1 Remote control

Large numbers of infrared emitting diodes (IREDs) are used in remote control applications. Well-known examples are remote control of TV sets and high-fidelity audio equipment, to change the program received and adjust volume, brightness, contrast, color, balance, tone and so on without leaving

one's arm-chair. Similar systems operated from a car can open
or close garage doors, or can be used as electronic keys
which are very difficult to falsify or copy, because the code
is complicated and can be changed frequently.

Because in this application the signal is to be transmitted
through the air without special measures for bundling, high
emitted powers and high sensitivity PIN detectors are required, whereas the speed of transmission is not particularly
high. Generally, pulse code modulation (PCM) methods are used
for remote control, which means the information to be transmitted is coded in pulse sequences with typical pulse dura-
tions of about 4 µs and repetition frequencies around 35 kHz.
Infrared emitting diodes on the basis of GaAs:Si or
$Ga_{1-x}Al_xAs:Si$ are well suited for this purpose. They emit
peak powers of 150 mW to 300 mW with driving pulses of 1.5 A.
IREDs for remote control applications are mainly packaged is
small angle, clear plastic cases as shown for visible LEDs in
Fig. 7-6. To increase the emitted power further, packages
containing two IRED chips are also available.

## 7-4-2 Optocouplers

Optocouplers are devices containing one emitter and one detector in the same case which are optically coupled, but
electrically isolated. Therefore they are sometimes also
called optoisolators. They can be considered as the smallest
possible complete optoelectronic systems. Optocouplers have a
large variety of applications when signals are to be transmitted from one part of a system to another, but when the
parts must remain electrically separated, for example to suppress interferences, to bridge high potential differences, or
for safety reasons. The most important parameters to describe
the properties of an optocoupler are the current transfer
ratio (CTR), the ratio of output to input current, which

characterizes the degree of coupling, and the maximum working
frequency which is a measure of speed. The CTR is influenced
by the efficiency of the emitter, the coupling between emitter and detector, and the sensitivity of the detector, whereas the maximum frequency is determined by the slower one of
the two components. As one might expect and as will be substantiated in the following paragraph, the two parameters CTR
and maximum frequency cannot be separately optimized. Instead, compromises must be found for different applications.

For frequencies up to about hundred kHz high current transfer
ratios over 500 % can be achieved using high-power IREDs on
the basis of GaAs:Si or $Ga_{1-x}Al_xAs$:Si and high-gain photo-
transistors. The use of the shorter wavelength (870 nm)
GaAlAs emitters (instead of GaAs (950 nm) results in an overall CTR increase of a factor of two. This gain is due to both
the higher power and the shorter wavelength of the GaAlAs
IREDs, because the sensitivity of silicon detectors decreases
rapidly for wavelengths in excess of 900 nm. For higher frequencies of several MHz considerably faster emitters and detectors have to be applied. Either infrared diodes based on
Zn-diffused gallium arsenide or visible LEDs consisting of
$GaAs_{1-x}P_x$ or $Ga_{1-x}Al_xAs$ can be used for fast couplers. Among
these, red $Ga_{1-x}Al_xAs$ LEDs exhibit a favourable combination
of high emitted power, short wavelength, and fast switching
behavior (see section 7-3-5). As detectors, pin diodes or
other high speed photodiodes come into consideration. Because
the CTR of a fast emitter/detector combination is quite
small, the detector signal has to be amplified, preferably
with an integrated receiver. For switching of high voltages
and high currents, couplers containing optically triggerable
triacs are also available, so called solid-state relays. Because these devices are mainly used for mains switching at
low frequency, high power infrared emitters can be applied
for triggering.

The geometrical configuration of optocouplers is dependent of the requirements to be fulfilled by the device. Some typical examples are shown in Fig. 7-11. Standard couplers are mostly of the coplanar type with emitter and detector on a common lead frame (Fig. 7-11, top). To increase the optical coupling, a reflector of white plastic is placed on top of the semiconductor chips. The interior of the reflector is filled with clear epoxy, and the whole device is molded with opaque plastic material. Even higher current transfer ratios can be achieved using a two-level configuration with the chips closely facing each other. This type of configuration suffers

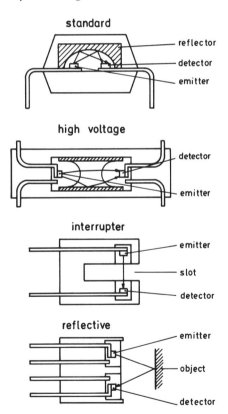

however from isolation problems and is more difficult to assemble. For high voltage application in particular the separation of the metal leads has to be increased (Fig. 7-11, second example). Emitter and detector are separately encapsulated and inserted into a short piece of white reflecting plastic tube. The metal leads are brought to the opposite ends of the outer case, so that isolation voltages in excess of 10 kV are achieved. The two remaining examples can be considered

Fig. 7-11 Different configurations for optocouplers, 1: standard type with coplanar configuration, 2: high voltage type, 3: interrupter, 4: reflective coupler

as optocouplers with an accessible optical path, which are used in many industrial applications. The interrupter (Fig. 7-11, third example) exhibits a slot where the radiation from the emitter to the detector can be interrupted by an object. This type of coupler is for example used for counting purposes. A reflective coupler (Fig. 7-11, bottom) features a parallel configuration of an emitter and a detector. If an object comes close enough to the device, part of the radiation is reflected back to the detector, so that this device can sense the presence or the approach of an object. Reflective couplers are very useful for automation purposes. A va-

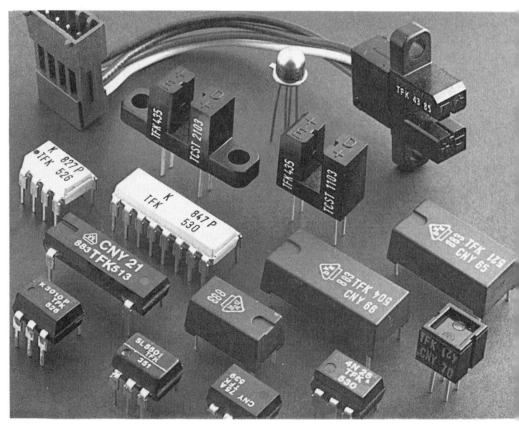

Fig. 7-12 Different optocouplers: standard and high
         voltage types (bottom and center), three
         interrupters and a metal can type (top),
         and reflective coupler (bottom right)
(Photograph courtesy by TELEFUNKEN electronic GmbH)

riety of couplers for different applications is shown in Fig.
7-12

## 7-4-3 Optical communication

Data transmission via optical fibers has many advantages over
electrical transmission: for example, lower weight and lower
price of the cable, very high bandwidth, potential separation, tapping safety, and immunity against interferences.
Depending on the requirements concerning, in particular,
transmission distance and bandwidth, different fibers, wavelengths, emitters, and detectors are used.

For long distances in excess of around 10 km and high bit
rates of several hundred Mbit/s transmission at 1.3 μm or
1.55 μm with quartz fibers, InGaAsP lasers and InGaAs detectors is suited best. In most countries this area of, for example, inter-city connections is served by public communication networks. For medium distances up to around 10 km and
lower data rates of the order of 100 Mbit/s., LEDs on the
basis of InGaAsP operating in the same wavelength range can
be used as transmitters. An example of this class of requirements is given by lines connecting single users to the
nearest telephone exchange. InGaAsP LEDs, which were developed for this purpose, are very similar to the corresponding
GaAlAs devices as described in section 4-2-3. For lower distances up to about 1 km transmission at shorter wavelength
around 850 nm through quartz fibers can also be applied. In
this wavelength range emitters based on $Ga_{1-x}Al_xAs$ and silicon detectors are used. High bit rates can again be achieved
only with lasers, whereas for lower bit rates in the range
from 10 Mbit/s to 100 Mbit/s surface or edge emitting LEDs
are applied (see section 4-2-3). Typical fields of application for this class of data links are local networks, factory
automation, and connections within ships or aircraft. If even

lower data rates of a few Mbit/s are to be transmitted over
very short distances of up to some 10 m, very inexpensive
systems consisting of visible GaAlAs LEDs emitting at 650 nm,
plastic fibers, and standard silicon detectors can be used
for optical communication (see section 7-3-5). Typical examples for this type of low cost applications are connections
within a house or from the central unit to a peripheral device of a computer system.

Optical communication is not only possible via glass or plastic fibers, but also through the air, although the unfavourable signal-to-noise ratio limits the application of this
method to transmission over a few meters within closed rooms,
for example for wireless sound transmission from a hi-fi unit
to headphones. Because for this application a bandwith of a
few 10 kHz is sufficient, high-power IREDs fabricated from
GaAs:Si or $Ga_{1-x}Al_xAs$:Si can be applied as transmitters in
this case.

Summarizing, it can be stated that infrared (sometimes also
visible) emitters can serve as transmitters for data communication at low to quite high data rates over short to medium
distances.

### 7-4-4 Further applications of IREDs

There are some further applications of IREDs which do not fit
into the categories of the preceding sections. Among these is
pumping of near-infrared solid state lasers, which was
achieved using very high power, double heterostructure LEDs
with hemisperical domes (ONO et al. 1976). Fast modulated
IREDs are also used for high precision distance measurement
systems, which rely on a phase difference measurement between
direct and reflected signal.

Finally, IREDs are applied in some autofocus systems, which are incorporated in many modern photographic cameras and slide projectors.

## Questions

Q7-1   What series resistor is required to operate an average LED with 10 mA from a 12 V source?

Q7-2   What has to be changed in the constant current circuit of Fig. 7-1 for an output current of about 10 mA?

Q7-3   Why is it not advisable to operate several LEDs in parallel?

## References

GILLESSEN, K., "A survey of interconnection methods which reduce the number of external connections for LED displays," Proceedings of the SID, 22/3 (1981), pp. 181-184

LEE, T.P. and T. LI, "Light-emitting-diode-based multimode lightwave systems," in Semiconductors and Semimetals, Vol. 22, Academic Press (1985), Orlando, Fla.

NAKAYA, S., K. NIHEI, S. KOTANI, I. ABIKO and A. NOMURA, "High-resolution display device using LED arrays," Proceedings of the SID, 23/3 (1982), pp. 197-202

NIINA, T., S. KURODA, H. YONEI and H. TAKESADA, "A high-brightness GaP green LED flat-panel device for character and TV display," IEEE Transactions on Electron Devices, ED-26 (1979), pp. 1182-1186

ONO, Y., M. MORIOKA, K. ITO, A. TACHIBANA and K. KURATA,
     "High Power LEDs for pumping of solid state laser,"
     Hitachi Review, 25 (1976), pp. 129-134

SAUL, R.H., "Recent advances in the performance and reliabi-
     lity on InGaAsP LED's for lightwave communcation sy-
     stems," IEEE Transactions on Electron Devices, ED-30
     (1983), pp. 285-295

SCHAIRER, W., "Properties of LED arrays for electrophoto-
     graphic applications", Journal of Imaging Technology, 12
     (1986), p. 76

YODOSHI, K., YAMAGUCHI, T., and NIINA, T., "A high-brightness
     monolithic display device using GaP green-light emitting
     diodes," Proceedings of the SID, 25/3 (1984), pp.
     201-205

## Answers to questions Q1-1 to Q1-6

Q1-1    4 (see Fig. 1-2)

Q1-2    There are 8 atoms in the unit cell, which has the volume $a^3$. Because the nearest neighbour distance is $a\sqrt{3}/4$, the radius of spheres is $a\sqrt{3}/8$. The volume of one sphere is therefore $4\pi/3 \, (a\sqrt{3}/8)^3 = a^3\pi\sqrt{3}/128$, and the volume of 8 spheres is $a^3\pi\sqrt{3}/16$. The fraction filled with atoms is consequently $\pi\sqrt{3}/16 \approx 34$ %.

Q1-3    The volume of the unit cell is $(0.5633 \text{ nm})^3 = 1.8065 \times 10^{-22} \text{ cm}^3$. There are 4 atoms of Ga or As in the unit cell, the requested number is therefore $2.2 \times 10^{22} \text{ cm}^{-3}$.

Q1-4    GaAs contains $2.2 \times 10^{22}$ As atoms per $\text{cm}^3$ (see Q1-3). $10^{18}/2.2 \times 10^{22} = 0.45 \times 10^{-4} = 0.0045$ %.

Q1-5    $(1/10^{-6} + 1/5 \times 10^{-8})^{-1}$ s   $4.8 \times 10^{-8}$ s = 48 ns

Q1-6    $D = \dfrac{kT}{e}\mu \quad \dfrac{1.38 \times 10^{-23} \text{ J K}^{-1} \times 300 \text{ K}}{1.6 \times 10^{-19} \text{ As}} \times 400 \text{ cm}^2/\text{V s}$

$\approx 104 \text{ cm}^2/\text{s}$

$$L = \sqrt{D\tau} = \sqrt{104 \text{ cm}^2/\text{s} \times 5 \times 10^{-8} \text{ s}} \approx 2.3 \times 10^{-3} \text{ cm} = 23 \text{ }\mu\text{m}$$

## Answers to questions Q2-1 to Q2-8

Q2-1    $\lambda$ (in nm) = $1240/E_g$(eV) = $1240/1.15$ = 1078 nm.

Q2-2    Because the radiative recombination is more efficient in these materials.

Q2-3    Boron nitride (BN) has the largest degree of similarity to diamond, because boron and nitrogen are direct neighbours of carbon in the periodic table of the elements. The same is true for indium antimonide (InSb) in relation to tin.

Q2-4    The properties, in particular the band gap energy, can be chosen according to the application.

Q2-5    Because most LED materials are grown epitaxially on a substrate crystal, and the lattice constants of the epitaxial layer must be approximately equal to the lattice constants of the substrate crystal.

Q2-6    0.557 nm. This result can be taken directly from Fig. 2-3, or it can calculated using VEGARD's law, i.e. by linear interpolation between 0.565 nm (GaAs) and 0.545 nm (GaP).

Q2-7    $E_g$ (in eV) = $1240/\lambda$ (in nm) = $1240/1378$ = 0.9 eV. The iso-bandgap line for 0.9 eV in Fig. 2-5 cuts the iso-lattice constant line for InP at x = 0.3, y = 0.35.

The required composition is therefore

$In_{0.7}Ga_{0.3}As_{0.65}P_{0.35}$.

Q2-8    It is the class of I-VII compounds, the alkali haloge-
        nides like sodium chloride, NaCl. These compounds have
        a purely ionic binding, whereas in III-V compounds the
        bonds are nearly covalent. With ionic binding the
        electrons are fixed to the atoms and do not give rise
        to electronic conduction.


## Answers to questions Q3-1 to Q3-5


Q3-1    The molecular weights of Ga, As and GaAs are 69.7,
        74.9 and 144.6, respectively. The amounts of Ga and As
        forming the crystal are therefore:

        Ga:    1000 g x 69.7/144.6 = 482 g
        As:    1000 g x 74.9/144.6 = 418 g

        To fill the ampoule (2 l) with 0.98 bar of $As_4$ vapor,
        n moles are necessary. From pV = nRT it follows that

$$n = \frac{pV}{RT} = \frac{9.6 \times 10^4 \text{ Pa} \times 2 \times 10^{-3} \text{ m}^3}{8.3 \text{ J/mol K} \times 883 \text{ K}} =$$

        $2.6 \times 10^{-2}$ mol

        This corresponds to $2.6 \times 10^{-2}$ x 4 x 74.9 = 7.8 g of
        $As_4$ molecules.
        Therefore 482 g Ga and (518 + 8) = 526 g As are re-
        quired.

Q3-2    The time $\tau$ necessary to diffuse along t is $\tau = t^2/D$
        which follows from $t = \sqrt{D\tau}$. The maximum cooling
        rate is given by MCR $= \Delta T/\tau$, the requested formula is
        therefore: MCR $= D\Delta T/t^2$.

Q3-3    1 $cm^2$ of a 1 mm thick Ga melt has a mass of 0.59 g
        corresponding to 8.5 x $10^{-3}$ moles. With a solubility
        of 1.8 at % this dissolves 1.8 x $10^{-2}$ x 8.5 x $10^{-3}$ =
        1.53 x $10^{-4}$ moles of P yielding the same number of
        moles of GaP. This corresponds to 1.53 x $10^{-4}$ x
        100.7 = 1.54 x $10^{-2}$ g of GaP. The volume of this
        amounts to 1.54 x $10^{-2}/4.13$ = 3.73 x $10^{-3}$ $cm^3$. With
        the area of 1 $cm^2$ the requested thickness is therefore
        3.73 x $10^{-3}$ cm or about 37 μm.

Q3-4    $C = \mathcal{E}_o \mathcal{E} \dfrac{F}{d}$ = 5.8 x $10^{-12}$ F

Q3-5    From Eq. (3-17):  $D = (x_j/2erfc^{-1}(N_b/N_o))^2 \dfrac{1}{t}$

        From Fig. (3-23):

        $x_j \approx$ 9.3 μm = 9.3 x $10^{-6}$ m at 730 °C = 1003 K

        $x_j \approx$ 3.3 μm = 3.3 x $10^{-6}$ m at 650 °C = 923 K

        t = 4 h = 14400 s

        The value of the inverse complementary error function
        $erfc^{-1}$ can be taken from Fig. 3-21 by interchanging
        the x and concentration axes. One obtains
        $erfc^{-1}$ $(10^{-3}) \approx$ 3.3 and $erfc^{-1}$ $(10^{-4}) \approx$ 3.8.

        Therefore D (1003 K) = (9.3 x $10^{-6}/2$ x $3.3)^2$ $\dfrac{1}{14000}$ $\dfrac{m^2}{s}$

                            = 1.38 x $10^{-16}$ $\dfrac{m^2}{sec}$ for $N_b/N_o$ = $10^{-3}$ and

$$D~(1003~K)~=~1.04~x~10^{-16}~\frac{m^2}{s}~\text{for}~N_b/N_o~=~10^{-4}$$

$$D~(~923~K)~=~1.74~x~10^{-17}~\frac{m^2}{s}~\text{for}~N_b/N_o~=~10^{-3}~\text{and}$$

$$D~(~923~K)~=~1.31~x~10^{-17}~\frac{m^2}{s}~\text{for}~N_b/N_o~=~10^{-4}$$

Using (3-18) the activation energy $E_a$ and $D_o$ can be calculated:

$$D(T_1)/D(T_2)~=~\exp\left\{-~\frac{E_a}{k}~(\frac{1}{T_1}~-~\frac{1}{T_2})\right\}$$

$$E_a~=~k~\frac{T_1~T_2}{T_1-T_2}~\ln~\left\{(D(T_1)/D(T_2)\right\}~;$$

$$k~=~0.834~x~10^{-4}~eV/K;~E_a~=~2.0~eV$$

$$D_o~=~D(T_1)/e^{-~E_a/kT_1}~=~3.34~x~10^{-6}~\frac{m^2}{s}$$

## Answers to questions Q4-1 to Q4-3

Q4-1    In LED types with an absorbing substrate the light
        emitted towards the substrate is lost. The reverse
        side contact can therefore be as simple as possible.
        With a transparent substrate, however, light passing
        the substrate can be utilized by providing a reflect-
        ing contact. Because contacts with good electrical
        properties are generally absorbing, the reverse side
        area is usually divided into a smaller well contacted
        and a larger reflecting part.

Q4-2    The two diode types differ in polarity: for the GaAs
        diode the substrate is the cathode in normal opera-
        tion, whereas the substrate is the anode for the
        GaAlAs diode.

Q4-3    An optimized edge emitter has a very thin active layer
        to reduce the internal absorption. Consequently the
        nonradiative interface recombination is increased as
        is the bandwidth. The lower output power is in part
        compensated by the narrower angle distribution of the
        output.

## Answers to questions Q5-1 to Q5-7

Q5-1    Definition of space angle $\Omega$: $\Omega = A/r^2$. Calculation of
        the surface area A of a sphere :

$$A = 2\pi \int_0^\alpha r^2 \sin\varphi \, d\varphi \quad , \text{ therefore}$$

$$x \text{ sr} = 2\pi \int_0^\alpha \sin\varphi \, d\varphi = 2\pi \left[ -\cos\varphi \right]_0^\alpha = 2\pi \left[ 1-\cos\alpha \right],$$

or $\cos\alpha = 1 - \dfrac{x}{2\pi}$ .

Solution $2\alpha$ (1 sr   ) = 65.5°

$2\alpha$ (0.1 sr ) = 20.5°

$2\alpha$ (0.01 sr) =  6.5°

Q5-2    a) $P_I = \int_\Omega J(\varphi) \, d\Omega = 2\pi \int_0^{\frac{\pi}{2}} J_0 \sin\varphi \, d\varphi = 2\pi J_0 \left[ -\cos\varphi \right]_0^{\frac{\pi}{2}}$

$= 2\pi J_0$

b) $P_L = 2\pi \int_0^{\frac{\pi}{2}} J_0 \cos\varphi \, \sin\varphi \, d\varphi = 2\pi J_0 \frac{1}{2} \left[ \sin^2\varphi \right]_0^{\frac{\pi}{2}} = \pi J_0$

Q5-3    $P_n = 2\pi J_0 \int_0^{\frac{\pi}{2}} \cos^n\varphi \, \sin\varphi \, d\varphi = 2\pi J_0 \left[ -\frac{1}{n+1} \cos^{n+1}\varphi \right]_0^{\frac{\pi}{2}}$

$= \frac{2\pi}{n+1} J_0$

Q5-4    $J(\varphi) = J_0 \cos^n \varphi_n = J_0/2$

$\cos \varphi_n = \frac{1}{\sqrt[n]{2}}$

$2\varphi_1 = 120°; \quad 2\varphi_2 = 90°; \quad 2\varphi_3 = 74.9°; \quad 2\varphi_4 = 65.5°;$

$2\varphi_6 = 47°.$

Q5-5    According to Fig. 5-4: color coordinates (x,y) approximately (0.52, 0.48), dominant wavelength 581 nm,
excitation purity 1.

Q5-6    Color coordinates (0.16, 0.61), dominant wavelength
516 nm, excitation purity 0.59.

Q5-7    Color coordinates (0.38, 0.28), dominant wavelength
not defined, because $(x_\lambda, y_\lambda)$ on purple line, excitation purity 0.29.

## Answers to questions Q6-1 to Q6-5

Q6-1   $P \sim I$,    $I_{SC} \gg I$

$I_F = I + I_{sc} \approx I_{sc} = A\left\{\exp\ (eV/2kT)-1\right\} \approx A\exp\ \dfrac{eV}{2kT}$

$P = A'I_0\left\{\exp\ (eV/kT)-1\right\} \approx A'I_0\ \exp\ eV/kT$

$P \sim I_F^{\,2}$

Q6-2   $\eta_i = \Delta n\ /\ \tau_r\ \left(\dfrac{\Delta n}{\tau_r} + \dfrac{\Delta n}{\tau_{nr}}\right) = 1/\tau_r\ \left(\dfrac{\tilde{\tau}_{nr} + \tilde{\tau}_r}{\tau_r\ \tau_{nr}}\right)$

$= \tilde{\tau}_{nr}\ /\ (\ \tilde{\tau}_{nr} + \tilde{\tau}_r)$

$\dfrac{1}{\Delta r} = \dfrac{1}{\Delta r_r + \Delta r_{nr}}$

$\dfrac{\tilde{\tau}}{\Delta n} = 1/\left(\dfrac{\Delta n}{\tilde{\tau}_r} + \dfrac{\Delta n}{\tilde{\tau}_{nr}}\right) = \dfrac{\tilde{\tau}_r\ \tilde{\tau}_{nr}}{\Delta n\ (\ \tilde{\tau}_{nr} + \tilde{\tau}_r)}$

$\dfrac{1}{\tilde{\tau}} = \dfrac{1}{\tilde{\tau}_{nr}} + \dfrac{1}{\tilde{\tau}_r}$

Q6-3   $\sin\alpha_T = 1\ /3.6 = 0.2\overline{7}$ ,        $\alpha_T = 16.1°$

$\sin\alpha_T = 1.5/3.6 = 0.41\overline{6}$,        $\alpha_T = 24.6°$

Q6-4    For air:

$$T(90°) = 4 \times 3.6/4.6^2 = 0.681, \quad T(\varphi) \leq T(90°)$$

$$f \leq T(90°) \int_0^{16.1°} I_0 \sin\varphi \, d\varphi \Big/ \int_0^{90°} I_0 \sin\varphi \, d\varphi$$

$$= T(90°) \left[ -\cos\varphi \right]_0^{16.1°} \Big/ \left[ -\cos\varphi \right]_0^{90°}$$

$$= T(90°) (1-0.961) = 0.0267$$

For epoxy:

$$T(90°) = 4 \times 1.5 \times 3.6/5.1^2 = 0.830$$

$$f \leq T(90°) \left[ -\cos\varphi \right]_0^{24.6°} \Big/ \left[ -\cos\varphi \right]_0^{90°}$$

$$= T(90°) (1-0.909) = 0.0753$$

Q6-5    $$P_0/I_0 \sqrt{1 + \omega_{max}^2 \tau^2} = P_0/2I_0$$

$$\omega_{max} = \sqrt{3}/\tau$$

## Answers to questions Q7-1 to Q7-3

Q7-1    $R = (12 \text{ V} - 2 \text{ V})/0.01 \text{ A} = 1000 \ \Omega$

Q7-2    The value of the 33 $\Omega$ resistor has to be doubled to
        yield the same voltage drop (the nearest value of

standardized resistors is 68 Ω). Because the output
current is reduced by factor of two, the required base
current for $t_2$ is also halved. The 10 kΩ resistor can
consequently be replaced by a 20 kΩ one.

Q7-3    If several LEDs are connected in parallel, they are
        operated at the same voltage. Even small differences
        in the forward characteristics cause considerable dif-
        ferences in the currents flowing through the indivi-
        dual diodes, because the forward current increases
        rapidly with increasing forward voltage.

# LIST OF SYMBOLS

a             lattice constant

A            area

B            recombination coefficient, luminance, brightness

$B_r$         radiative recombination constant

c            velocity of light ($2.9975 \times 10^8$ m/s)

$c_n$         Auger coefficient for n type material)

$c_p$         Auger coefficient for p type material)

d            thickness

D            diffusion constant, irradiance

e            elementary charge ($1.6022 \times 10^{-19}$ As)

$E_a$         activation energy, acceptor energy

$E_b$         band bending

$E_c$         energy of conduction band

$E_d$         donor energy

$E_g$         band gap energy

$E_t$         trap energy

$E_v$         energy of valence band

f            modulation bandwidth

             fraction of radiation that escapes through a surface

F            luminous flux

$g_o$            generation rate, thermal equilibrium

$h$             Planck constant ($6.6262 \times 10^{-34}$ J s)

$h,k,l$      Miller indices

$i$             current density

$i_o$           static current density

$i_1$           time dependent current density

$I$             flux density, luminous intensity, current

$I_o$           saturation current

$I_{sc}$         space charge recombination current

$j$             imaginary unit, $j^2 = -1$

$J$             radiant intensity

$k$             Boltzmann constant ($1.3807 \times 10^{-23}$ J/K)
                wavevector, distribution coefficient

$K$            Photometric radiation equivalent for photopic vision
                (673 lm/W)

$K'$          Photometric radiation equivalent for scotopic vision
                1725 lm'/W).

$L$             diffusion length of minority carriers, illuminance

$L_n$          diffusion length of electrons

$L_p$          diffusion length of holes

$n$             number, numeric factor, refractive index

$n_e$          number of electrons

$n_i$          intrinsic electron concentration

$n_p$          number of photons

$N$             concentration

$p$             hole concentration

$P$             radiant flux

$P_v$          spectral irradiance

| | |
|---|---|
| r | radius, distance, recombination rate |
| $r_r$ | radiative recombination rate |
| $r_{nr}$ | nonradiative recombination rate |
| $r_o$ | recombination rate, thermal equilibrium |
| R | radiance, resistance, reflectivity |
| | |
| s | interfacial recombination velocity |
| | |
| t | thickness, time |
| T | absolute temperature, transmissivity |
| | |
| V | voltage |
| | relative luminosity function for photopic vision |
| V' | relative luminosity function for scotopic vision |
| $V_g$ | voltage corresponding to the band gap energy |
| $V_{pn}$ | voltage across the pn junction |
| $V_T$ | temperature voltage (kT/e) |
| | |
| w | width |
| | |
| x | mole fraction in mixed crystals, distance |
| | coordinate in the CIE chromacity diagram |
| | |
| y | mole fraction in mixed crystals |
| | coordinate in the CIE chromacity diagram |
| | |
| $\bar{x}, \bar{y}, \bar{z}$ | color matching functions |
| | |
| $\alpha$ | angle |
| | absorption coefficient |
| $\alpha_T$ | angle of total reflection |
| | |
| $\Delta$ | spectral halfwidth |
| $\Delta r$ | difference between total recombination rates |

$\Delta r_{nr}$    difference between nonradiative recombination rates
$\Delta r_r$    difference between radiative recombination rates
$\Delta t$    time interval

$\mathcal{E}$    excitation purity
$\mathcal{E}_o$    permittivity of vacuum ($8.8542 \times 10^{-12}$ F/m)
$\mathcal{E}_s$    static dielectric constant

$\eta$    yield, efficiency
$\eta_c$    escape probability
$\eta_{ext}$    external efficiency
$\eta_i$    internal efficiency
$\eta_I$    injection efficiency
$\eta_p$    power efficiency
$\eta_v$    photometric efficiency
$\eta_s$    sputter yield

$\lambda$    wavelength
$\lambda_d$    dominant wavelength

$\mu$    mobility
$\mu_e$    electron mobility
$\mu_a$    hole mobility

$\nu$    frequency

$\tau$    minority carrier lifetime
$\tau_r$    radiative lifetime
$\tau_{nr}$    nonradiative lifetime

$\varphi$    angle

$\emptyset$    potential
$\emptyset_b$    barrier height
$\emptyset_F$    Fermi level

$\emptyset_s$          potential difference

$\omega$           circular frequency  $\omega = 2\pi f$

$\Omega$           space angle